International Rainwater Catchment Systems Experiences

Towards Water Security

International Rainwater Catchment Systems Experiences

Towards Water Security

Edited by

José Arturo Gleason Espíndola, César Augusto Casiano Flores, Raul Pacheco-Vega and Margarita Rosa Pacheco Montes

IWA PUBLISHING

Published by **IWA Publishing**
Alliance House
12 Caxton Street
London SW1H 0QS, UK
Telephone: +44 (0)20 7654 5500
Fax: +44 (0)20 7654 5555
Email: publications@iwap.co.uk
Web: www.iwapublishing.com

First published 2020

British Library Cataloguing in Publication Data
A CIP catalogue record for this book is available from the British Library

ISBN: 9781789060577 (paperback)
ISBN: 9781789060584 (eBook)

Contents

Chapter 2
Harvesting the potential

Chapter 3
Transforming tradition of rainwater harvesting in Sri Lanka

About the Authors

Chapter 1
Han Heijnen – Vice-president of the International Rainwater Catchment Systems Association, consultant in water issues in many parts of the world. He has worked with WHO for many years and is president of the International Rainwater Harvesting Alliance (IRHA) based in Geneva.

Chapter 2
Billy Kniffen – National trainer in the American Rainwater Catchment Systems Association (ARCSA). Retired professor at A&M Texas University and Abeline University in Texas. He founded the Texan Rainwater Catchment Systems Association. His work is well known around USA and Latin America.

Chapter 3
Tanuja Ariyananda – Chief Executive Officer/Director of Lanka Rain Water Harvesting Forum (LRWHF) since its inception in 1996. LRWHF is a non-government not-for-profit organisation working to promote Rainwater Harvesting for domestic use in Sri Lanka. Hydrobiologist with a PhD in Environmental Science. Member of the Ministerial Committee for Developing Policy and Strategy for Rain Water Harvesting in Sri Lanka 2005 Past President (2008–2012), Secretary General (2003–2008), Director Women's programme (2001–2003) of International Rainwater Catchments Systems Association. Author of several books on rainwater harvesting, and published more than 25 research papers at national and international forums.

Chapter 4
Patricia Macomber – University Outreach Educator, created and ran a highly successful water quality and quantity program for the state of Hawaii, which

addressed the needs and concerns of people using rainwater catchment systems. In 2001, she authored the nation's first statewide '*Guidelines on Rainwater Catchment Systems for Hawaii*' which was adopted by the Hawaii State Department of Health and the US Veteran's home loan program.

Chapter 5

Jose Arturo Gleason Espíndola – Researcher at University of Guadalajara, Founder and President of Mexican Rainwater Catchment Systems Association (AMSCALL, initials in Spanish). He is also director of Technological Water Research Institute Lic. Arturo Gleason Santana A.C. He is regional director in Latin America of the International Rainwater Catchment Systems Association.

Yahaira Corona Sánchez – Project Coordinator at the Technological Water Research Institute Lic. Arturo Gleason Santana A.C. Member of the Mexican Rainwater Harvesting System Association.

Cesar Casiano Flores has a Ph.D. in Innovation and Governance for Sustainable Development, University of Twente. Currently, he is postdoctoral researcher at KU Leuven and affiliated researcher at University of Twente and Universidad Iberoamericana campus Puebla. He is member of the National Researchers System in Mexico (SNI-CONACYT) and qualified university lecturer in The Netherlands (UTQ/BKO). He has coordinated different environmental and research projects worldwide. He has more than 10 years of experience in education and research in Mexico and Europe. His current research is focused on water sensitive cities and land governance issues.

Chapter 6

Margarita Pacheco – Founder of the International Rainwater Harvesting Alliance (IRHA) in 2002. She lived in Switzerland for long time leading IRHA. Now an international consultant in water affairs and is working for Colombian government.

Chapter 7

Makoto Murase – Rainwater Catchment Pioneer, winner of Rolex Award for Enterprise 2016. Author of the book '*Rainwater and you*' that was presented in Spanish during the 1st RWH Mexican Conference, November 2017.

Chapter 8

David Vargas – Founder of Isla Urbana. He studied civil engineering at Pennsylvania State University.

Enrique Lominitz – Founder of Isla Urbana (NGO) that has lead the rainwater cause across Mexico.

Chapter 9
Anne Kikundwa – Liaison Officer, works with Uganda Rainwater Association. She has experience in managing projects with a background in social work and social administration. 10 years work experience in the water and sanitation sector, of which seven years were at programme level and three years in managerial positions for water and sanitation specialised organisations (Kigezi Diocese Water and Sanitation Programme, and Uganda Rainwater Association).

Chapter 10
Sene Souleymane – Rain Foundation in Casamance and near Dakar. He has developed social projects through the Caritas association in Africa.

Chapter 11
Macpherson Nthara – Twelve years of progressive work and responsible experience in land use planning, and the promotion of sustainable land management i.e. rainwater harvesting, soil and water conservation, flood mitigation and climate change adaptation projects.

Chapter 12
Verónica Livier Díaz Nuñez – Researcher at University of Guadalajara. She is an expert in urban planning and founding member of AMSCALL.

Chapter 13
Prof. Mooyoung Han – Department of Civil and Environmental Engineering at Seoul National University. He is now a Fellow of IWA and the chairman of IWA RWHM specialist group since 2003. He received 2010 and 2012 PIA awards with the promotion of rain cities. He is doing voluntary service of rainwater harvesting demonstration projects in Korea and developing countries with many innovative ideas. Recently, he authored '*Hydrological Design of Multipurpose Micro-catchment Rainwater Management*' published by IWA.

Chapter 14
Bob Boulware – President of DAE and Principal in Charge. Past National President, American Rainwater Catchment Systems Association (ARCSA); Life Member, American Society Heating Refrigeration, Air Conditioning Engineers (ASHRAE); Chapter President, National Codes and Standard Committee for American Society of Plumbing Engineers (ASPE).

Chapter 15
Catalina C. Dobre – Investigates the transition of urban areas towards sustainable or water sensitive environments. Her work provides new insights into sustainable water management by exploring participatory evaluation methods in urban planning and design.

Chapter 16

Johann Gnadlinger – Founder and President of Brazilian Rainwater Catchment System Association. He was the General Director of the International Rainwater Catchment Systems Association.

Chapter 17

Marco Schmidt – Professor at Technical University of Berlin. He is member of scientific group called 'Watergy'. He has done a lot of rainwater harvesting work in Germany and other countries around the world.

Chapter 18

Paul Akkerman – worked in different jobs in industry and vocational education in The Netherlands. He has worked in rural development work in Guinea-Bissau, West Africa, in industry and vocational education in The Netherlands. Also he is funder and project leader of independent RWH project in Guinea-Bissau, named CLEAN WATER – HEALTHY VILLAGE (De Gevulde Waterkruik). Now he is an independent consultant and trainer for domestic RWH systems in different countries of East and West Africa.

Preface

The scarcity of water will be one of the most critical problems in the world in the next 15 years as a result of population growth and contamination. Also, the effects of climate change will worsen the environmental situation with intensive droughts on one hand, and damaging floods on the other. Rainwater harvesting (RWH) has emerged as a critical strategy to meet these challenges through the experience and knowledge of teams of international experts. Since Agenda XXI (1992), RWH is considered as a significant action to meet water scarcity particularly in islands and in the expansion of the water supply network (United Nations Sustainable Development, 1992).

Furthermore, the United Nations launched the Millennium Development Goals Report in 2015. In this document the water issue is included in goal number six that states, 'By 2030, ensure availability and sustainable management of water and sanitation for all' (United Nations, 2018). In referring to RWH, this goal proposes increasing the efficiency of irrigation (more crop per drop) and harvesting and reusing rainwater at the household level in water-poor regions.

The rainwater harvesting cause needs more promotion throughout the world. In spite of the advances around the globe, in many countries it is not a well known technology. For example, Latin America and some Asian countries would benefit greatly from building on the experience available from countries. This book promotes different skills, knowledge, and approaches from several experts who have worked in various professional fields in several countries. You will find

 International Rainwater Catchment Systems Experiences: Towards Water Security
Editor(s): José Arturo Gleason Espíndola, César Augusto Casiano Flores, Raul Pacheco-Vega
and Margarita Rosa Pacheco Montes
doi: 10.2166/9781789060584_xix

experts with tremendous experience and expertise alongside emerging experts with enthusiasm and new ideas.

The book has four sections: basic concepts, narratives of RWH, programs implemented by diverse sectors of society, and notable technical cases. The first part explains a short history of RWH and the functioning of its components, with purpose of giving the reader the basics. It allows the reader to obtain a variety of approaches, examples, knowledge that comes from the daily life of each author. Given the importance of rainwater as a tool to tackle the current and future water scarcity problems of the world, this document provides an overview of the state of the art of RWH advances from many contexts across the planet.

The subject of this book is related to the promotion of different international rainwater experiences that provides sustainable water services and climate resilience, including technical aspects and socio-cultural and policy affairs. This book is the result of the efforts of various international experts who have been driving rainwater harvesting systems as a real alternative for supplying water and counteracting climate change effects over the last 30 years. This volume appeals to a wide range of readers who are interested in the rainwater harvesting being a useful tool in engineering, architecture, and urbanism, programs. The reader can gain the inspiration with the aim of getting involved in the international rainwater catchment cause through knowing and contacting many experts with a vast professional experience in different disciplines and approaches.

Several of the contributions are derived from the First National Conference organized by Mexican Rainwater Catchment Systems Association (AMSCALL, initials in Spanish), International Rainwater Catchment Systems Association (IRCSA), University of Guadalajara, and the National Council of Science and Technology from Mexico, celebrated in 2017 in Guadalajara, Jalisco, México.

I would like to thank every contributor who worked with dedication and commitment; your effort is not in vain, you are planting a seed that will grow significantly for the benefit of humanity and the planet.

Dr. Arturo Gleason
Book Editor
President of the Mexican Rainwater Harvesting Systems Association (AMSCALL)

REFERENCES

United Nations (2018). Sustainable Development Goals, Knowlegde Plataform. Recuperado el 29 de November de 2019, de https://sustainabledevelopment.un.org/sdg6

United Nations Sustainable Development (1992). Agenda 21. Recuperado el 29 de November de 2019, de https://sustainabledevelopment.un.org/content/documents/Agenda21.pdf

Introduction

Water determines where people can live. Water is crucial to all human needs and most activities. It allows cities to grow, land to be fertilized and economic activities to develop. Water is used for agriculture, for life and for energy. In Roman times water was carried through aquaducts which were strictly controlled for leakage and proper use. Rivers and catchments have been protected to ensure sufficient and clean water. Water is stored during rainy periods to tide over the dry season. Over time, people became engineers to manage their living environment.

When there was no water anymore, people moved their cattle, their businesses, to places where water could be secured, sometimes seasonally, sometimes forever. With growing populations, fewer opportunities to move to new lands, and with challenging climate and topography, better water solutions and sharing are needed to allow societies to prosper.

It is nice to talk about those who have access to good water supply. But there are many households and communities that live in areas that are limited in water resources, where water may only come during seasonal rains. Women may have to travel a distance to collect water for the home. Agriculture revolves around periods of rainfall.

Small island communities, fishing villages along the coast or areas with saline ground water wells often find it hard to find fresh water for drinking and cooking. Sometimes the groundwater is contaminated with arsenic or fluoride, making its consumption hazardous to health.

In upland areas, where households are situated above the springs, women have to trek arduously to collect a reasonable amount of water for their families.

Editor(s): José Arturo Gleason Espíndola, César Augusto Casiano Flores, Raul Pacheco-Vega
and Margarita Rosa Pacheco Montes
doi: 10.2166/9781789060584_xxi

Even in cities, there are many examples of where water supply has not yet reached the whole population as utility companies struggle to keep up with the growth and demand.

Dying Wisdom, a book about traditional water harvesting in South Asia, states that 'water harvesting emerges as a practice related to local community needs and sensitive to local ecological demands' (CSE, 1999). It describes the experience and expertise of ordinary people able to manage their livelihood in areas with limited rainfall.

This IWA publication presents a number of rainwater harvesting initiatives from around the world where practitioners have demonstrated that rainwater harvesting makes a vital difference for a lot of households and communities. Better rainwater-harvesting techniques for domestic water, including drinking water, and for use in small scale agriculture makes a great difference to all lives, women and men, young and old, the able and less able.

The WHO-UNICEF Joint Monitoring Programme for Water Supply and Sanitation reported that in 2017 90% of the world population had access to safely managed water or a basic service with safe water (JMP, 2019). That leaves 10% of the world population without a decent water service!

Harvesting rain and using it for drinking and domestic use will help to give scattered households and underserved city dwellers access to water. The world cannot ignore using rain as the source of all water. National and local rainwater harvesting policies and strategies need to be applied to make sure that best use can be made of rain to provide water to individual households, to store or buffer water in community reservoirs and to augment the aquifer using infiltration techniques. Not all of these experiences are described in this book, as this publication has a greater focus on domestic retention and use of rainwater. Nevertheless, climate change effects have gradually forced cities to think about slowing down the water that falls on its territory and make sure it does not go to waste. The Sponge City concept and similar approaches are gradually demonstrating their value and longer-term cost saving. More recently, growing emphasis on greening the city and reducing heat stress near buildings offers further beneficial applications of rainwater harvesting. Educating people, and especially young people, on the value of water, and the need to use rainwater at least once before letting it go, will lead to better water management.

Alert planners increasingly recognize that convenience and health can also be achieved with decentralized services. The case studies in this publication demonstrate this. By incorporating rainwater harvesting in the portfolio of water supply solutions, governments will meet 'the human right to water [is] indispensable for leading a life in human dignity'. Everyone recognizes the right to water. One does not deny drinking water to a guest or passing stranger. Many cultures demonstrate this by first bringing a glass of water when one is invited into the house.

The UN Committee on Economic, Social and Cultural Rights defines the right to water as the 'right of everyone to sufficient, safe, acceptable and physically accessible and affordable water for personal and domestic uses'. The UN Sustainable Development Goals underline this more specifically by stating in SDG 6.1 that the 2030 objective is to achieve universal access to safely managed water: in effect water from an improved source, available in the premises, when it is needed and safe to drink. Governments have an obligation to implement the human right to water and ensure that universal access is achieved (Albuquerque & Roaf, 2014).

Safely harvested rainwater when properly stored largely fulfills the SDG 6.1 criteria on safely managed water. In addition, rainwater contributes to several more SDGs, in agriculture, poverty alleviation, gender, water resource management, ecosystem protection, and so on.

This book presents stories and experiences from some 15 countries from all over the globe, developed and less developed. There are many more experiences that can be highlighted, but these are thought to give the reader a good overview of what can be done when water supply engineering investment has not yet reached the unserved. Rainwater harvesting has an important role to play as a safe water supply when other sources are not safe or not available. A rainwater system requires storage. This may be somewhat expensive but will most often be cheaper than a house connection from a centrally treated water supply system. This storage could be very valuable in an emergency situation and in times of drought when households with rainwater storage will still have access to water.

We encourage you, the reader, to peruse these stories and arguments at your leisure. There are many ideas and techniques to be gleaned that will be applicable in planning for the serving of the last 10% who are still waiting for water security and a good water service.

Enjoy the enthusiasm and the humanity behind the stories.

REFERENCES

Albuquerque C. De. and Roaf V. (2014). *Realizing the Human Rights to Water and Sanitation* (Vol. 1). Human Rights to Water & Sanitation UN Special Rapporteur, New York.

CSE (1999, Chapter 4). Fourth Citizens Report SOE-4: Dying Wisdom: Rise, Fall and Potential of India's Traditional Water Harvesting Systems. A. Agarwal and S. Narain (eds.), Centre for Science and Environment (CSE), New Delhi.

JMP (2019). *Progress on Household Drinking Water, Sanitation and Hygiene 2000–2017*. Special Focus on Inequalities, United Nations Children's Fund (UNICEF) and World Health Organization, New York.

Section 1

Basic Concepts

I. The importance of rainwater catchment systems

J. A. Gleason Espíndola[1] and Y. Corona Sánchez[2]

[1]University of Guadalajara, Guadalajara, Mexico (Arturo.Gleason@cuaad.udg.mx)

[2]Technological Water Research Institute Lic. Arturo Gleason Santana A.C, Guadalajara, Mexico (yahaira.corona@iitaac.org.mx)

I.1 History of rainwater catchment systems

Rainwater harvesting (RWH) has been present in cultures since ancient times, as a vital activity that required specific technical expertise and social organization. Harvesting rain for domestic or agriculture was a way of life. There are records that in the ancient city of Byblos in Lebanon approximately 4900 years ago, it had an efficient sewage system, storage, and catchment of rainwater (Fernández Pérez, 2009). By the year 1000 B.C. on the Arabian Peninsula, in the upper areas of Yemen, buildings and temples had courtyards and terraces that facilitated the catchment and storage of scarce rainwater (Ballen *et al.*, 2006).

The Roman culture implemented functional rainwater harvesting systems in their houses (Fernández Pérez, 2009). In the early stages, the Romans used a pool close to the springs named *lacus*, when these pools where underground they were called cisterns. Later, in the *domus* or Roman house it was articulated around inner courtyards. The central *atrium* was open to the sky, where the rainwater from the roofs was guided to a central pool (*impluvium*) for use and storage in the house. As a result of urbanization and the growing population, the consumption of water increased. This led to the development of covered cisterns. These cisterns were

 International Rainwater Catchment Systems Experiences: Towards Water Security
Editor(s): José Arturo Gleason Espíndola, César Augusto Casiano Flores, Raul Pacheco-Vega
and Margarita Rosa Pacheco Montes
doi: 10.2166/9781789060584_0003

built in the ground underneath the courts with two significant advantages: firstly, the amount of water which could be stored increased considerably and evaporation losses reduced; and secondly, the cisterns served as a protection against water pollution (Ballen *et al.*, 2006).

The order of Caesar Justinian constructed one of the biggest cisterns of the world with a capacity of 80,000 cubic meters and measures of 140 by 70 meters. After the Roman Empire fell and with the arisal of new cultures like the Muslim culture, new ways of life and water management emerged.

In the arid regions of Iran, agricultural and permanent settlements were supported by the ancient *qanat* system of tapping alluvial aquifers at the heads of valleys and conducting water along underground tunnels by gravity, often over many kilometers (UNESCO, 2016).

Each qanat comprises an almost horizontal tunnel collecting water from an underground water source, usually an alluvial fan, into which a mother well is sunk to the appropriate level of the aquifer. Well shafts are sunk at regular intervals along the route of the tunnel to enable removal of spoil and allow ventilation. These appear as craters from above, following the line of the qanat from a water source to agricultural settlement. The water is transported along underground tunnels, called *koshkan*, utilizing gravity with the gentle slope of the tunnel to the exit (*mazhar*), from where channels distribute the water to the agricultural land of the shareholders (UNESCO, 2016).

In ancient Israel, the genius of human creativity is represented in Masada by its sophisticated water system. This system transformed a barren, isolated natural fortress set in an arid, dry climate (less than 50 mm yearly rainfall and without natural resources) into a lavish royal retreat with grand, classic bathhouse, two large swimming pools, and ample water supply. Masada's system utilized run-off water from one rainy day to sustain life for up 1,000 people for 2–3 years (UNESCO, 2000).

This water system was so accurately designed and executed to capture run-off water in such a desert climate. On the mountaintop, gutters and canals diverted water from roofs and unbuilt areas into small pools and grand cisterns. These cisterns also received run-off water from flash floods of riverbeds west of Masada.

Approximately 2000 years ago, an elaborate water harvesting system developed in the hinterland of the ancient capital Anuradhapura in Sri Lanka. This system allowed the collection, storage, and distribution of rainfall and run-off to supply a growing population with food under the semi-arid climatic conditions. The *Wewas* are reservoirs and the cornerstone of a highly sophisticated water harvesting system.

I.2 Rainwater harvesting in Mesoamerica

In Mesoamerica, water symbolized the generator of the development of local civilizations. The establishment of the first civilizations with seasonal agriculture made the cultures in Mesoamerica masters of irrigation techniques, including rain and stormwater irrigation systems, as is evidenced in various archeological investigations. The Mesoamerican cultures took advantage of the rainwater through natural or artificial ways, where they captured and retained water for the population.

The Mayan culture was a great manager of its water through the use of *Chultuns* in all Mayan regions, and was able to capture and distribute rainwater for agricultural and human consumption. Chultuns were underground excavations waterproofed with gesso used since the pre-classic period (Ballén Suárez *et al.*, 2006).

In some cities of the Olmec culture can be found the oldest vestige of rain- and stormwater management in Mesoamerica: La Venta (now Tabasco, Mexico) used sewers and stone channels, and San Lorenzo (Veracruz, Mexico) used mud pipes and underground aqueducts of carved basaltic stone, joined with a mix of chapopote (Rojas, 2009).

On the other hand, the city of Tenochtitlan, founded in 1342, was not only the greatest achievement of Mesoamerican hydraulic engineering, but one of the greatest in the history of ancient man. The city constituted itself as a challenge of political and social organization, where the catchment of rainwater was vital for the operation of a higher self-sustainable city. Tenochtitlan was founded on a basin with a system composed of lakes, rivers, lagoons, and swamps, fed rain, permanent and semi-permanent rivers, as well as springs. Tenochtitlan was surrounded by water all the time. It required an overwhelming effort of hydraulic engineering and colossal construction work to sustain and maintain the necessary hydraulic infrastructure (Gutierrez, 2014). With the arrival of the Spanish conquerors, these techniques were complemented and in many cities replaced the traditional pre-Columbian rainwater harvesting systems.

The Spanish techniques introduced by the conquest allowed better excavation of cisterns and deeper wells with the use of the pulley, the wheel and the lathe. This resulted in a more efficient extraction of water. Colonial houses had pools as a means to contain the rainwater, while sanctuaries, convents and churches built reservoirs to store the water driven by pipes from the roofs (Rojas, 2009, p. 20 and 22).

Rainwater harvesting in barrels, pots, reservoirs, and wells was a typical system, which lasted in some places until the beginning of the 1970s, such as Merida, Zacatecas, and Guanajuato. However, in the current situation of worldwide water scarcity, it would be worth re-examining old techniques and taking advantage of what new water management and rainwater harvesting techniques bring to achieve sustainable communities and cities.

Throughout the Andes, the Incas built a series of dams that, by their location, are called Altoandinas, whose function was to store rainwater and then use it during the dry season. The Amunas is an ancestral practice to recharge aquifers. This practice consisted of a ritual that involved assemblies, cleaning of ditches, and, above all, reverence of water.

The system worked with open ditches that followed contours lines, allowing the rainwater to gather in cochas (pools), which were open, and then the water filtered into the mountain.

The water emerged downstream as puquios (underground aquaducts) months later and during the dry season, when sewage was more marked in the basin, they were used for agriculture. This kept the people provided with food (Ancajima, 2014).

I.3 Recent efforts of rainwater harvesting around the world

For thousands of years, collecting rainwater was a standard method for providing water. Over the last century, wells and municipal water supplies took over as primary water sources. However, in the past 40 years, there have been several attempts to create a new rainwater culture around the world. In the early 1980s, a pioneer in rainwater harvesting, Makoto Murase, designed a water recovery system. The system collects, filters, and stores rainwater in sizeable underground holding tanks, easing flooded sewers and providing a resource used for irrigation, toilets, washing, and drinking (Strauss, 2016).

Nowadays, over a thousand Tokyo buildings harvest and recycle rainwater, since the system installed in Tokyo's Sumo Stadium proved so successful that the city eventually required underground rainwater tanks for all new buildings.

In 1989, after the *4th International Rainwater Cistern Systems Conference* in Manila, the International Rainwater Catchment System (IRCSA) was founded. IRCSA aims to promote rainwater catchment systems planning, development, management, science, technology, research, and education worldwide and has established an international forum for scientists, engineers, educators, administrators, and those concerned in this field. IRCSA drafts international guidelines on this technology and updates and disseminates information, as well as collaborating with and supporting international programs (IRCSA, 2013).

After the formation of the international foundation, different countries started to create Rainwater Associations to promote sustainable rainwater harvesting practices, such as the American Rainwater Catchment Systems Association (ARCSA). ARCSA efforts include: creating a favorable regulatory atmosphere, creating a resource pool and educating professionals and the general public regarding safe drinking rainwater design, installation, and maintenance practices.

Every year communities all around the world are attempting to cope with water scarcity and floods. Therefore, to promote rainwater harvesting as an effective and sustainable solution for water shortages and to prevent natural hazards, including floods and droughts from weakening the resilience of communities and the ecosystems upon which they depend, in 2002, The International Rainwater Harvesting Alliance (IRHA) was established in Geneva. IRHA develops projects that reinforce the resilience of communities and restore ecosystems through better management of rainfall and run-off (IRHA, 2017).

Climate change projections raise concerns over changes in temperatures and rainfall patterns and, ultimately, over water supply to communities. Historically, following periods of drought or water shortage, rainwater is beneficial to provide drinking water to communities all around the world. In Australia, the Millennium drought left cities such as Melbourne a year away from running out of the water with extremely low dam levels. The drought was broken by a period of intense rainfall, where cities like Brisbane suffered major floods that caused billions of dollars worth of damage. Townsville suffered extreme monsoon conditions with dams reaching more than 200% capacity and triggering floodgates to open, causing flooding for thousands of residents. Therefore, Australia is transforming urban areas to ensure longterm liveability, sustainability, and resilience in the face of climate change and population growth through adaptative infrastructure, innovative and strong economies (Rogers & Hammer, 2019).

Urban Rainwater harvesting in those affected regions has dual benefits, supplementing municipal water supply and the potential to improve urban stream hydrology by capturing, consuming, and effectively removing excess urban run-off (Taylor & Brodie, 2016). Rainwater tanks are encouraged and being installed in urban areas, resulting in an increase resilience of cities to droughts and a reduction of mains water demand (van der Sterren *et al.*, 2012).

Rainwater harvesting for domestic use is a prevalent practice in islands such as Thailand, the Caribbean, Hawaii, etc., particularly in areas not served by municipal water. In 2010, an estimated 30,000 to 60,000 people in Hawaii were dependent on rainwater catchment systems to satisfy their water needs (Donohue *et al.*, 2017).

II. FUNDAMENTALS OF RAINWATER CATCHMENT SYSTEM

Rainwater harvesting is a technique of collection and storage of rainwater into natural reservoirs or tanks, or the infiltration of surface water into subsurface aquifers (before it is lost as surface runoff). One method of rainwater harvesting is rooftop harvesting. With rooftop harvesting, almost any surface (tiles, metal sheets, plastics, but not grass or palm leaf) can be used to catch the flow of rainwater and provide a household with high-quality drinking water all year-round. Other uses include water for gardens, livestock, irrigation, and industrial supply.

A rainwater harvesting/catchment system consists of seven phases for proper functioning (See Figure S1.1):

(1) Catchment area

The catchment area is generally roofs, patios, garages, asphalted roads, or any non-permeable surface where rainwater flows and collection is feasible. Catchment areas need to have a slope that allows the water to flow directly to the gutters or downspouts for its conveyance.

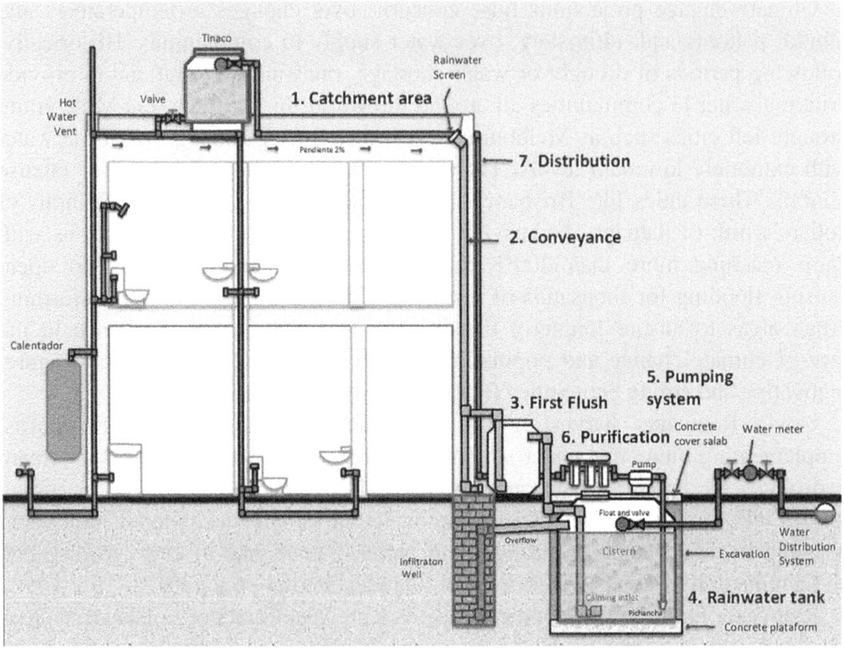

Figure S1.1 Rainwater Catchment Systems Components. *Source*: Authors.

The catchment area can be composed of many materials such as cement, metal, fiberglass, plastic, roof tiles, etc. Some elements can generate water losses or release toxins, so it is noteworthy to consider the material for the catchment area.

(2) Conveyance system

The conveyance system includes the set of gutters or pipes that drive rainwater from the catchment area to the storage system. Gutters and downspouts must be in accordance with the requirements of the system and the characteristics of the building. The roof, gutters, and downspouts should be durable, sized correctly, and accurately placed in the building where they will be used (Kniffen, 2012).

(3) First flush

The first flush diverter is designed to hold the water from the first rains of the season. The first flush diverter keeps the first flow of poor quality roof runoff from entering the tank. Diverters must have a drainage outlet for emptying standing water and be emptied as needed. This device is especially needed when trees and birds are close to the catchment area and depends on the catchment area size.

Additionally, to keep the water clean, prevent clogging, and sediment build-up, basic filtration is required. Before the water goes to the storage tank, it is necessary for a leaf screen, so the water inside the tank is as clean as possible.

(4) Water storage tank

The water storage is the equipment that stores and protects the water. A water tank needs to consider price, size, aesthetics, and/or water use. The tank capacity is determined by supply and demand.

Rainwater tanks need at least three connections: (a) a water inlet with optional turbulence calming device to prevent remixing of sediment; (b) an overflow inlet located below the water entrance level; (c) pipe to the pump and pressure tank; and optionally a connection to municipal water/make up water.

Considerations for rainwater tanks should include the following: above ground-tanks should be UV resistant to prevent sun damage; mosquito and animal-proof tank-access points – microbial insecticide specific to mosquito larvae can be used; overflow should be directed to a useful area, away from tank foundations, buildings, and toward plants or another tank; and any tank used for underground storage must be rated for this use.

(5) Pump system

The pumping system distributes the water from the storage tank to the places of use. It should be noted that the suction pipe of the pump must be at least 50 cm above the bottom of the tank to prevent sediments (Palacios, 2010). The pump needs to have enough power to send water through the filters and to the place of use.

(6) Filtration

This is the purification process to which rainwater is subjected to guarantee a certain quality, so it meets the necessary conditions for its use. The purification process depends on water use. When rainwater is used for human consumption, three filters are required: polypropylene, carbon block, and UV light.

(7) Distribution

Once the rainwater has the quality needed for its required use, it can be distributed to where it is needed. Commonly, the rainwater distribution system is separated from the municipal water supply system. Therefore, there will be two distribution networks. Usually, the rainwater distribution pipes are painted purple in color to differentiate between networks. The rainwater is consumed for different purposes, mainly human, domestic, industrial or agricultural uses.

This book aims to present different experiences of systems installed at homes, buildings, and roofs. To learn more about other forms of rainwater harvesting and use, the following are recommended reading.

- *Harvest the Rain. How to enrich your life by seeing every storm as a resource*, Nate Downey (Ed) (2010) Sunstone Press, Santa Fe, New Mexico.

- *Rainwater Harvesting Manual* 1st edition, Ann Audrey (Ed) (2015) ARCSA, Tempe, Arizona.
- *Rainwater Tank Systems for Urban Water Supply. Design, yield, energy, health risks, economics, and social perception*, Ashok K. Sharma, Donald Begbie & Ted Gardner (2015) IWA, London, UK.
- *Designing Rainwater Harvesting Systems. Integrating rainwater into building systems*. Celeste Allen Novak, Edward Van Giesen & Kathy M. DeBusk (2014) Wiley, New York.

REFERENCES

Ancajima R. (2014). Tecnologías Ancestrales-sistemas Hidráulicos pre Incas e Incas. Ministerio del Ambiente, Lima.

Ballén Suárez J. A., Galarza García M. and Ortíz Mosquera R. O. (2006). Historia de los sistemas de Aprovechamiento de agua de lluvia. IV SEREA – Seminario Iberoamericano sobre Sistemas de Abastecimiento Urbano de Agua. João Pessoa, Brasil.

Ballen, Galarza and Ortíz (2006). Historia de los Sistemas de Aprovechamiento de agua de Lluvia. Seminario Iberoamericano sobre Sistemas de Abastecimiento Urbano de agua Joa Pessoa (págs. 1–12). Joao Pessoa, Brasil: Universidade Federal de Paraíba.

Donohue M., Macomber P., Okimoto D. and Lerner D. (2017). Survey of Rainwater Catchment Use and Practices on Hawaii Island. *Journal of Contemporary Water Research & Education*, **161**(1), 33–47.

Fernández Pérez I. (2009). Aprovechamiento de Aguas Pluviales. Escola Politécnica Superior de Edificación de Barcelona, Cataluña, España.

Gutierrez A. (2014). Captación de agua pluvial, una solución ancestral. En Impluvium Sistemas de Captación de Agua de Lluvia (págs. 6–11). Periodico digital de divulgación de la Red del Agua UNAM, Mexico.

IRCSA (21 de March de 2013). IRCSA. Obtenido de International Rainwater Catchment System Assoaciation: http://ircsa.in/

IRHA (2017). IRHA. Obtenido de International Ranwater Harvesting Alliance: https://www.irha-h2o.org/en

Kniffen B., Clayton B., Kingman D. and Jaber F. et al. (2012). *Raiwater Harvesting: Planning process*. Texas A&M AgriLife Extension Service. College Station TX, U.S.A.

Rogers B. and Hammer K. (2019). Realising the vision of a water sensitive city. The Source. Obtenido de https://www.thesourcemagazine.org/realising-the-vision-of-a-water-sensitive-city/

Rojas T. (2009). Semblanza Histórica del Agua en México. SEMARNAT Secretaría de Medio Ambiente y Recursos Naturales, Mexico.

Strauss G. (13 de November de 2016). Reining in the Rain. Obtenido de National Geographic: https://www.nationalgeographic.com/news/2016/11/makoto-murase-explorer-moments-rain-water-conservation-Japan/

Taylor B. and Brodie I. (2016). Rainwater harvesting in Australia for Water Supply and Urban Stream Restoration. Enviromental benefits of urban rainwater harvesting. IWA World Water Congress & Exhibition, Brisbane.

UNESCO (2000). Masada Proposed World Heritage Site. UNESCO, Israel.

UNESCO (2016). UNESCO. Obtenido de The Persian Qanat: https://whc.unesco.org/en/list/1506/

van der Sterren M., Rahman A. and Dennis G. (2012). Rainwater harvesting systems in Australia. Ecological Water Quality-Water Treatment and Reuse in Australia (págs. 471–496), En D. Voudouris (ed.), InTech. Obtenido de https://www.intechopen.com/books/ecological-water-quality-water-treatment-and-reuse/rainwater-harvesting-systems-in-australia

Section 2

Narratives of RWH

Chapter 1

Rain: Water for health, livelihood and self-reliance

H. A. Heijnen

International Rainwater Harvesting Alliance (IRHA), Geneva, Switzerland
han.heijnen@irha-h2o.org

Keywords: drinking water, rainwater harvesting, safely managed water supply, SDG6

1.1 INTRODUCTION

The WHO UNICEF Joint Monitoring Report on the *Progress on household drinking water, sanitation and hygiene 2000–2017* states that 90% of the world population has access basic or safely managed water supply. By 2030 global coverage will be around 96%, falling short of the universal access agreed under the Sustainable Development Goals (JMP, 2019). Water supply is deemed to be safely managed if the water is provided through an improved water supply source, available on the doorstep, anytime when needed, and safe to drink. Properly constructed and maintained treated piped-water supply, boreholes, protected springs, protected dug well and domestic rainwater harvesting are technologies that qualify.

To date some 10% of the world population is still lacking a decent water supply service. As all the 'easy' water supply services have been provided, servicing the last 10% is getting harder because of issues of a technical, water resource, social or financial nature. However, SDG6 stipulates universal access and solutions need to be found to develop adequate services, even if these are somewhat more

Editor(s): José Arturo Gleason Espíndola, César Augusto Casiano Flores, Raul Pacheco-Vega and Margarita Rosa Pacheco Montes
doi: 10.2166/9781789060584_0015

expensive than usual. Note that the observations made in this chapter will implicitly refer to low and lower middle income countries where the provision and extension of services such as water supply is often financially, socially and technically challenging.

While most people would prefer a centrally serviced piped water supply reaching their home, this is not always possible. Remoteness, topography or very low population density may make such a supply too costly. Globally 64% of the population enjoys access to a piped supply (JMP, 2019). The rest of the population has a protected dug well on the premises or a rainwater harvesting system, or has to collect water from community standposts, wells or boreholes. Those who have no such facility nearby are forced to collect their water from a stream or pond. Clearly water quality and convenience diminish rapidly through this summary.

In many parts of the world, central supplies now also face seasonality of water sources, forcing reduced consumption during several dry months. Households try to overcome such periods of shortage by managing their consumption, but also by employing additional water sources. For instance, regular droughts have encouraged Australian households to install rainwater harvesting systems in urban homes, making the water for all last longer. In Sri Lanka, many homes in the Western and South Western part of the island have protected dug wells which can be used to back up the central supply or even, with a small electrical pump, fill up a roof-level storage tank and facilitate internal water supply. This is as good as a centrally managed piped supply.

The effects of climate change are gradually being noticed everywhere. Dry periods are lasting longer and when it rains, it comes in shorter, intense showers. Run-off is high due to dry soils and to urban paving and the water that is needed to bridge the dry season disappears quickly. Climate change forces society to change its habits and adjust its physical planning to ensure local and regional water security while managing reduce flood risks.

1.2 HARVESTING RAIN, GOING GLOBAL

From time immemorial people have been managing rain. Availability of water and water sources determined where people would be able to live. Adequate rainfall decided on the quality of pasture or the ability to grow grain. Technical advances and finance may have enabled societies to inhabit big cities and expand agriculture into dry areas, but only because of the resource rain provided through the water cycle.

Today, some 1.3% of the world's population uses rainwater as its main source of domestic water. In developing countries, this may be up to 2.4% of the rural population (JMP, 2017a). In Australia 26% of households use rainwater at the home, while for 23% of those living outside capital cities it is a common source of drinking water (Australian Bureau of Statistics, 2013).

Where water has been traditionally scarce or difficult to retain due to a karstic geology, as in Jamaica or parts of the Adriatic Coast, people have found ways to collect and store rainwater. As a city-state, Singapore manages the rainwater that falls on its territory to the full, to raise its water security. Since the mid-nineties, Germany has taken the lead to decouple rainwater from the sewerage system, initially to postpone the costs of renewing the sewerage systems in its growing towns by reducing run-off peakflows, but later also to develop safer solutions to use the rainwater mostly for non-potable purposes in the home. The appreciation of the potential to use rainwater harvesting and reduce flooding risks has led many countries to implement policies to encourage better management of rain water in their cities. These initiatives go by various names, but the Sponge City concept in China is evocative as it creates the mental image of absorptive capacity in the city during rain events and subsequent use of the retained water in drier periods. Encouraging use of rainwater as a complementary source of water makes a difference to the household and the city.

This book contains narratives of domestic rainwater harvesting initiatives in Asia (Bangladesh, Japan, Korea and Sri Lanka), Africa (Guinea Bissau, Malawi, Senegal and Uganda) and the Americas (Brazil, Colombia, Hawaii, Mexico and the USA). It is but a selection of successful programmes and leaves out, for instance, Thailand and Kenya, both with substantial rainwater harvesting expertise and continued progress. Thailand was an early adopter in the 1970s and ‘80s, developing more affordable storage reservoirs for the dry North East of the country. Today it still claims 15% national domestic RWH, rising to 23% for the rural areas (JMP, 2017a).

The interest in rainwater harvesting grew in the 1980’s when it became clear that important areas of the world would start facing water stresses that would reduce their capacity to achieve food security. The International Rainwater Catchment Systems Association (IRCSA) was launched in 1991 as an outcome of the *4th International Rainwater Cistern Systems Conference* of 1989 in Manila. Its membership was mainly academic and for some three decades it informed the water development sector through regular scientific conferences. Through the IRCSA conferences practitioners around the world learned from the work of Prof. Zhu Qiang of the Gansu Institute of Water Conservancy, China, and have adapted this experience to overcome their problems from Africa to North Eastern Brazil.

The interaction with the International Rainwater Catchment Systems Association (IRCSA) at the 2nd Brazilian Rainwater Catchment Symposium, in July 1999, demonstrated to Brazil the importance and potential of rainwater harvesting. The Conference led to the creation of ABCMAC: Associação Brasileira de Captação e Manejo de Água de Chuva (Brazilian Association for Rain Water Harvesting and Management). Since then, Brazil has launched the P1MC domestic RWH programme: Programa um Milhão de Cisternais rurais (the One Million Rural Tank programme) in the vast dry North East of the country. It was later reinforced by a programme that augmented rainwater harvesting for agriculture as well: P1+2: Programa uma Terra e duas Águas (Programme One Piece of Land and Two Sources of Water) (Heijnen, 2013). So far the two programmes report to have facilitated the construction of some 730,000 rainwater harvesting systems in homes, farms and schools (asabrasil.org.br). Really getting towards the one million systems!

In the water thirsty world of today RWH can support livelihood through rainfed agriculture. This is practiced in a very high percentage of farm lands: 95% in Sub-Saharan Africa; 90% in Latin America; 75% in the Near East and North Africa; 65% in East Asia; and 60% in South Asia (IWMI, 2010). The World Agroforestry Center in Kenya builds on this by promoting the Billion Dollar Business Alliance for Rainwater Harvesting towards food and water security urging farmers in Sub-Saharan Africa to construct farm ponds and harvest water during the rainy season (kenyacic.org/news/billion-dollar-business-alliance). Kenya and Ethiopia are known for the development and the use of sand dams in dry rivers, a system that captures and stores the occasional bursts of rain. In addition, domestic rainwater harvesting is spreading, with Kenya nationwide having 5.4% of users.

Advocacy for rainwater harvesting has branched out to argue for improving the health of watersheds through retention and slowing down the run-off of water to encourage restoration of aquifers through managed aquifer recharge and so coax wells to gradually become perennial again, e.g. in India, Ethiopia, China.

Following recommendations formulated during the *World Summit for Sustainable Development* in Johannesburg in November 2002, the International Rainwater Harvesting Alliance was established to advocate and inform about rainwater harvesting. It is mainly active in support of regional and national rainwater organizations in developing countries (irha-h2o.org/0).

The interest in domestic rainwater harvesting in countries like Australia, Germany and the United States has financed research and product development with respect to optimization of filters to remove dirt e.g. Leaf Beater Downspout Filter; first flush diverters; storage tanks and rainwater harvesting fittings, etc. In Western Europe and the UK, municipal governments, researchers and entrepreneurs are increasingly engaged in bringing local and domestic rainwater harvesting within the water resources management continuum (Melville-Shreeve *et al.*, 2016).

The German Ministry of the Environment states on its website on the use of rainwater in the household that '*the technology and products for using rainwater in the home have now matured. Technical standards are available … However, rainwater systems need regular check-up and care. This is the duty of the user*'.

Now that the world is facing the effects of climate change, requiring ever more water for its cities and committing itself to the Sustainable Development Goals of providing, among other, all people with a safe water source for drinking and personal hygiene, it is worth examining rainwater harvesting as an instrument to help extend a safely managed water supply service.

1.3 RAINWATER HARVESTING FOR DOMESTIC WATER SUPPLY

Rainwater can be harvested as a domestic supply, for use as drinking water and for food preparation, as well as for a variety of other household chores. In addition, rainwater and wastewater may be used for maintaining a vegetable plot, some small livestock and productive trees.

1.3.1 Why collect rainwater for drinking and domestic needs?

Where water supply is not available, not sufficient or safe to drink, harvesting and storing rainwater may be part of the solution. Households living in such areas are already frugal with water and will not allow it to go to waste. Efficient water use comes with lifestyle, for personal hygiene, ecological sanitation or gardening with mulching and planting drought resistant crops.

1.3.1.1 In semi-arid areas or at an uphill elevation

Retention, storage and conservation of rainwater is a matter of survival, water security and self-reliance, for the household and for (minor) agriculture, to tide over the dry period.

1.3.1.2 When ground water has a bad taste

In coastal areas and small islands, but also in some inland aquifers, water can be saline and not fit for drinking. Groundwater may have a high iron or manganese content that people may find objectionable and is difficult to remove.

1.3.1.3 Substitution

Where boreholes yield water with excess levels of arsenic or fluoride, or in situations where people suffer from chronic kidney disease of uncertain etiology (CKDu) as in Sri Lanka or Central America, consumption of ground water would eventually lead to serious health consequences. Consumption of rainwater would be a safe alternative.

1.3.1.4 Rainwater may bring a better service

In urban areas where water supply is under stress and costly, rainwater harvesting will augment the existing service, reducing consumption from the public supply and assuring greater water security at home. Uncoupling the rainwater downpipe from the city sewer and slowly draining the rainwater on unpaved premises allows for infiltration improving the aquifer.

Figure 1.1 Leaflet from Lanka Rainwater Harvesting forum (Dissanayake, 1996).

1.3.2 Managing the stored water

In terms of quantity, there is sufficient water during the rainy periods, but when the rain stops management of the rainwater store is needed. It has often been a reason for planners to disregard rainwater harvesting as it is not reliable in terms of a perennial supply.

However, for the households that practice rainwater harvesting this is hardly ever a concern. With rain, the household will have several months of excellent service during the rainy periods, while in the intervening months or the dry season the use of rainwater should be managed for quality purposes, as it is normally a safe source, or for accessibility, e.g. for a system serving an elderly couple.

People in vulnerable households already manage their water from several sources depending on the season. The women have an understanding of the availability, quality and preference of water sources and have learned to live with it. However, the installation of a rainwater harvesting system raises water security and ease of access at least for drinking water. The storage that a rainwater system brings, would also allow for a water tanker to fill up the reservoir during the dry period. Even in urban areas, harvesting rainwater can very well enhance service levels by augmentation of a city supply that is insufficient or erratic.

1.3.3 Quality of harvested and stored rain water

The quality of the rainwater collected and stored depends on the cleanliness of the catchment area (roof); the application of a first flush to divert the first, potentially dirtier water; the use of an entry filter and the inlet–outlet system used in the tank (Pathak & Heijnen, 2007). When properly managed, rainwater is of good quality and mostly better than public supplies that are inadequately treated (World Health Organization, 2017).

Epidemiological studies show that consumers are not at greater microbial risk than users of piped systems (Heyworth *et al.*, 2006). Nowadays good and affordable household filters are easily available to make the collected rainwater completely safe. When lead flashings and roofing paint are not used, chemical risks are also minimal (enHealth, 2010).

In 2017, a *Review of Roof Harvested Rainwater in Australia* looked at 148 studies and reported on the use of rainwater harvesting in Australia from the perspective of chemical and microbial health. It concluded that chemical constituents were generally within accepted health guidelines, except where due to past mining or industry activities dust blown onto the catchment area may cause exceedences. Rainwater can easily be microbiologically contaminated if the first flush is not used properly. Even then, the review stated that the epidemiological evidence in Australia does not point to consumers being affected by drinking untreated rainwater. At the same time, the study advised for household water treatment to ensure the water is bacteriologically safe (Chubaka

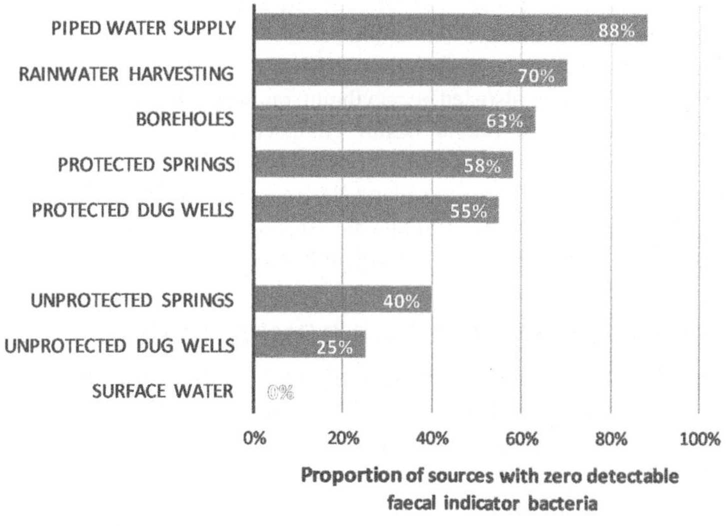

Figure 1.2 Comparative water quality realized among technologies. Studies from low and middle income countries by R. Bain *et al.* (WHO, 2017).

et al., 2018). Note that when water is stored in the tank for a while, as will usually be the case, the water quality will improve due to die-off of bacteria in a low-nutrition environment and the effect of biofilm on the inside of the tank. The biofilm will also absorb heavy metals (Evans *et al.*, 2009).

Figure 1.2 shows a comparison between technologies in use in low and middle income countries against their safety as a drinking water source. It shows that in 70% of cases rainwater is a safe source even when it usually not treated before drinking (JMP, 2017b).

1.3.4 Initiating rainwater harvesting programmes

When alternative water supply solutions are not available, rainwater harvesting implementation programmes should be considered. Domestic rainwater harvesting systems are obviously household based and this may lead government and agencies to consider that investment costs should be borne by the householder. However, getting a connection to a piped scheme in the city or the installation of a borehole in the community are most often subsidized. As the water storage tank is expensive, building your own system may be impossible without government support or a bank loan.

So, from a perspective of equity and promoting universal access to safe water supply, government support for rainwater harvesting system development would be justified. Moreso, as in most developing countries the beneficiaries of the rainwater systems would mostly be the poorer sections of society.

The large majority of rainwater harvesting systems (as described in this book) are directly subsidized by the government or its agents. In the case of developing countries, sponsors may provide non-governmental organizations (NGOs) financial support to undertake the implementation. In Australia application of rainwater harvesting is encouraged or mandated through a mix of subsidy and rules. That situation has also led to a vibrant rainwater harvesting market with associated regulatory and public health guidance (Chubaka *et al.*, 2018). In Germany and the Netherlands, the focus has been more on decoupling of rainwater from the sewer line and using the rainwater at home for flushing or gardening, but also here, households are encouraged to take the first steps through attractive offers and subsidy through the municipality.

1.4 USE OF RAINWATER FOR DOMESTIC USE

Data have been obtained from the WHO/UNICEF Joint Monitoring Programme for Water Supply, Sanitation and Hygiene (JMP) in late 2017 to make an assessment of how far rainwater is currently being used as a source of drinking water. On the basis of the data received, 98 countries were included in the sample. Member countries of the Organization for Economic Co-operation and Development (OECD), except Australia and Mexico and Russia were not included in the sample. Data are from the updated 2015 reporting by JMP, as used for the publication on 'Progress on

Drinking Water, Sanitation and Hygiene: 2017 update and SDG baselines' (JMP, 2017a).

Countries that reported less than 0.1% usage of rainwater as a drinking water supply or with data from before 2010 are excluded from the sample. This resulted in 115 countries, of which again 17 were eliminated as data were not available or unclear. As a result, most countries included are from Latin America, Africa and Asia. That is not to say that there are no rainwater users in the Northern hemisphere, but the data are as yet not available and users are not likely to use rainwater for drinking because their economies and infrastructure are more sophisticated.

The JMP lists rainwater harvesting as an improved water source, however as it is supposed to be a small contributor to the goal of universal water supply, it is at this time not making a separate assessment of rainwater use in the world.

The data do show some interesting results. Presently some 86 million people (say 17 million households) state in surveys that they are using rainwater for human consumption, or 1.6% of the 5.3 billion people included in the sample (or 72% of the world population). Considering only the rural population, the usage figure goes up to 2.4% of the 2.8 billion rural population.

Obviously many more collect rainwater for other domestic purposes. The data from Australia indicate that more than twice as many households use collected rainwater for other purposes around the home. If that would be more generally true, 3 to 3.5% of the population living in non-OECD countries and Russia would be familiar with rain-water harvesting.

It is likely that we see an up-tick in the number of households in Europe that will adopt rainwater harvesting at least for non-potable applications. In Australia this already the case, as Figures 1.3 and 1.4 show.

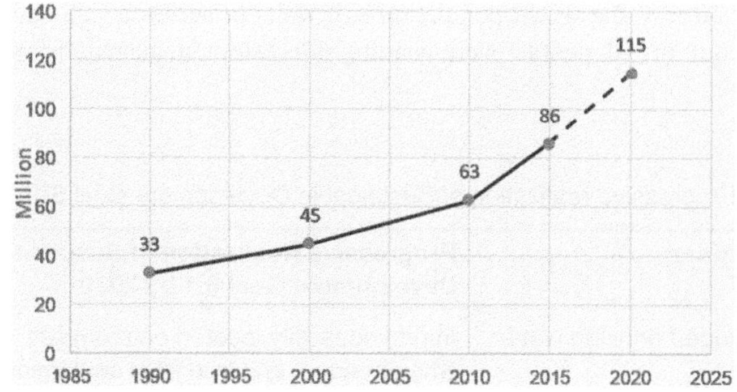

Figure 1.3 Growing use of domestic rainwater harvesting for drinking water in 2015 (JMP, 2017a), without OECD countries and Russia, incl. Australia and Mexico, in million people served.

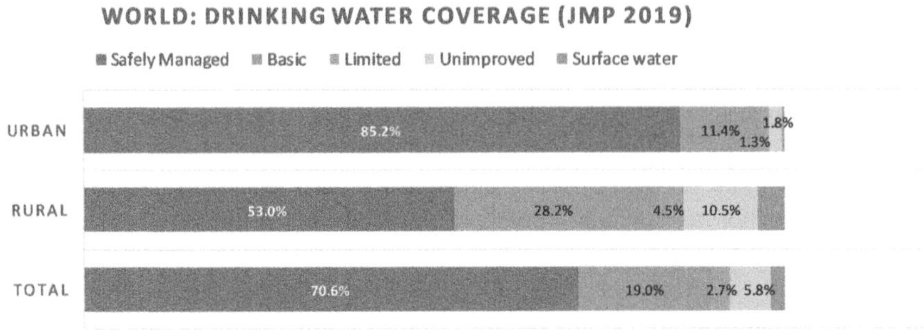

Figure 1.4 WHO/UNICEF JMP (2019) Progress on household drinking water, sanitation and hygiene 2000–2017. Special focus on inequalities.

1.5 SUSTAINABLE DEVELOPMENT GOAL (SDG) 6

Harvested rainwater can play an important role in improved water supply, in particular support for a well constructed and operated system meets the criteria for a safely managed supply set forth by SDG 6.1: '*By 2030, achieve universal and equitable access to safe and affordable drinking water for all*'.

Until 2015, the Millennium Development Goals were in use as the international targets. These stipulated relative simple and clear services in water supply using an improved source and in sanitation for safe removal of faeces. For hygiene, handwashing with water was sufficient to meet the standard. The adoption of the 2030 Agenda for Sustainable Development changed the baseline for WASH.

New, more stringent 'technical' indicators for WASH-related Sustainable Development Goals have been introduced. Table 1.1 'provision of safe drinking water' this means that: the water supply tap point should be near the home, in effect on the premises; it should be available when it is required (from a city-network tap, or a storage tank); and it should be free from contamination.

In 2017 10% of the world population still waits to access a safe water supply. Overall 3 out of 10 people were waiting for safely managed drinking water

Table 1.1 Progressive realization of Sustainable Development Goal 6.1.

Service Ladder	Progressive Realization of Sustainable Development Goal 6.1 by 2030
Safely managed drinking water supply	Improved facility located on premises, available when needed, and free from contamination
Basic water supply: continuation of MDGs	Use of improved water source, within 30 minutes round trip collection time

Source: JMP (2017b)

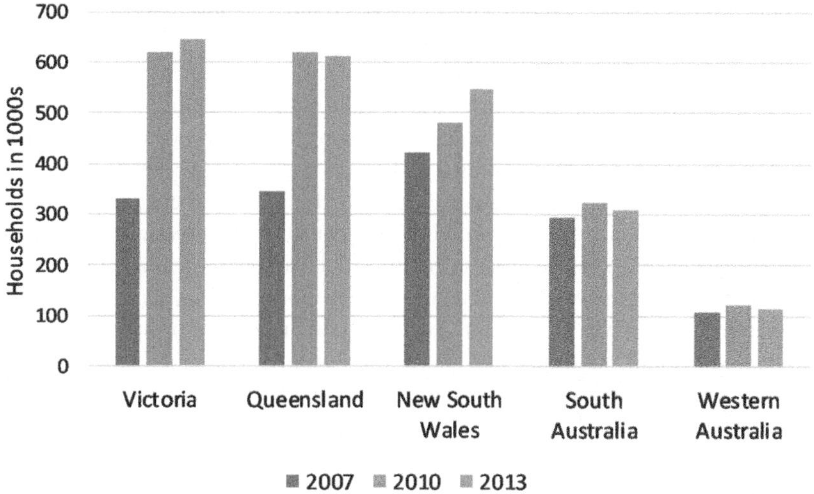

Figure 1.5 WHO/UNICEF (JMP, 2017a) Progress on household drinking water, sanitation and hygiene 2000–2017. Special focus on inequalities.

services in 2015. Those that are unserved are mostly in low and lower middle income countries, and mostly in Figure 1.5.

In achieving the SDGs, greater emphasis is placed on realizing equity of access and inclusion. This is also expressed home in Target 1.4: 'Achieving universal access to basic services', which aims to ensure all that men and women, in particular the poor and vulnerable, have equal rights to economic resources, as well as access to basic services [including WASH]. Other SDG6 targets raise the importance of water efficiency, integrated water resources management and ecosystems.

1.6 HOW TO PROVIDE WATER SUPPLY SERVICES TO THE LAST 10% OF THE UNSERVED?

How can we tackle the remaining SDG water supply challenges? The straightforward, easier and less costly water supply projects have been completed. And even these now face maintenance problems, upgrading and water shortage episodes, so more work is needed to keep these in good shape.

From now on, all water supply projects will be more difficult as water requirements by different parties at local, regional and national level require more negotiation, technical and hydrological know-how, policy agreement on water resources management, etc. Not all the unserved communities and households will be affected by these constraints as they may have traditionally been relying on local sources and agreements. But those who live in the periphery of cities experience shortages every day, as the cities grow and the capacity of water utilities to provide basic water services does not keep pace.

In addition to the 10% of the world population still without an acceptable service, 19% have a basic service. Apart from water quality, water quantity is a concern for these two categories. Additional alternative sources may be needed to meet the increase in demand. In urban settings, additional water sources need to be found or secured (at times with ecosystem charges to be paid to the guardians of the source area), to ensure that the growing city can provide adequate water to its citizens. To manage such a process requires a clear policy, public information campaigns, and extensive communications with consumers. Clarity of purpose and enabling municipal policies in Australia convinced authorities to invest in additional rainwater solutions that work for everyone, do not break the bank (even though the solutions may require a substantial financial commitment) and enhance water security all round.

What options are available to provide a service to the last 10%? Technical solutions such as desalination are expensive to run, thus, would attract a subsidized environment, which would not be sustainable. Similarly, lift-pumped supplies to serve households in hill areas do not come cheap in recurrent charges either. So how long can a rural community manage without having to beg support from the local government? Diverting water from another river basin is only for the desperate, and at a serious cost. Also, the cost of transporting the water is clearly adding a substantial recurrent expense due to the depreciation of the pipeline and the cost of pumping charges. Thus households can resort to bringing water by tanker. This is costly and may well be tricky with respect to quality and regulation.

1.7 BE READY TO COLLECT AND STORE RAIN

A functioning, well-managed rainwater harvesting system fulfills the SDG 6.1 criteria nearly in full: it provides a water supply service on the premises, it is available anytime and the water quality is very good. If in doubt, one can treat the water through a household water treatment system. Rainwater harvesting faces the problem that the quantity of water may not be adequate all year round, but that could be overcome by increasing storage if rainfall and roof collection area allow this.

Rainwater harvesting meets several other SDG criteria. It will fulfill the right to water for the household and lead to sharing of water collection tasks in the household. The reduced time spending on daily water collection will afford a woman opportunities to engage socially and raise her status by taking on some income generating productive work, for instance in growing and selling vegetables. Kitchen farming will improve nutritional intake, health is better protected and children have less water carrying tasks, so they will be in school on time. A rainwater tank in the yard will also satisfy the needs of elderly people or persons living with a disability (Jones, 2013).

Through rainwater harvesting, users will develop a better appreciation of climate change and this will probably raise interest and capacity to improve household and community climate change resilience. At any rate, water security in the home will improve substantially. It may well be that water will be scarce during the dry season, but women have proven to be good managers using various strategies to keep water in the rainwater tank longer. If all else fails, and if the road allows, water can be tankered in to make up for a shortfall. In Hambantota, a drought prone southern district in Sri Lanka, those with rainwater tanks, do usually not need a top up with tankered water distributed by the local administration. People manage.

It is true that climate change has created new challenges. Historical precipitation data are no longer that reliable, as rainfall has become more erratic. Delays of the onset of the expected rainfall create problems for farmers sowing crops or transplanting rice seedlings. When the rains come, it is more often a burst of water overwhelming the collection systems. Key words are erosion control, ponding and storage, managed aquifer recharge, restoration of wetlands and promotion of viable ecosystems, managed for water conservation, etc. Research, piloting and effective monitoring will help to develop the most practical solutions for affordable, incrementally upgradable rainwater harvesting systems that will serve households and institutions such as schools and health care facilities, and ensure that rain is increasingly seen as the invaluable resource that it is.

Depending on local conditions, rainfall and local capacity and political interest, it should now be possible to accelerate the uptake of rainwater harvesting solutions for domestic and community water supply in situations where otherwise, for technical and management reasons, a service is not feasible. The technology is, with different levels of technical sophistication, increasingly available in the market and it will be possible to adopt solutions that will work. To serve those who reside in remote, dispersed hamlets, on hill sides, or in the un(der)served periphery and slums of cities governments should be prepared to provide additional financial and technical support to reach out to the last unserved.

Rainwater is a valuable resource, which should be exploited in the most efficient way. Governments should explicitly recognize rainwater harvesting in their water supply policy to encourage and promote rainwater harvesting for human consumption and domestic use, to protect the people's health and livelihood, and ensure good utilization and conservation of water resources.

1.8 CONCLUSION

Currently, RWH for domestic use and for human consumption is a reality for many households. It is estimated that around 6–8% of the world population could be served substantially through improved RWH as a drinking water source and for other domestic usage.

Where rainwater harvesting is feasible, it addresses the human right to a water supply service, can provide a safely managed service with limited extra costs, frees women from drudgery and raises their independence (Rautanen *et al.*, 2015). It improves water security, and sanitation and hygiene in the household.

Without proper use of rainwater, universal access to safely managed water supply as defined under the Sustainable Development Goal 6 will not be achieved.

REFERENCES

Australian Bureau of Statistics. (2013). Environmental Issues: Water use and Conservation, Mar 2013. Retrieved from https://www.abs.gov.au/ausstats/abs@.nsf/Lookup/4602.0.55.003main+features3Mar2013

Chubaka C. E., Whiley H., Edwards J. W. and Ross K. E. (2018). A review of roof harvested rainwater in Australia. *Journal of Environmental and Public Health*, **2018**, 6471324, https://doi.org/10.1155/2018/6471324

Coleman, S. (2016). *Built environment: Urban environmental efficiency: Water efficiency. Australia state of the environment 2016.* https://soe.environment.gov.au/theme/built-environment/topic/2016/urban-environmental-efficiency-water-efficiency

enHealth. (2010). Guidance on the use of rainwater tanks. Retrieved from https://www.sa.gov.au/__data/assets/pdf_file/0012/14142/WSUD_chapter_5.pdf

Dissanayake, H. (1996). *Promoting Rainwater Harvesting in Sri Lanka* (p. 4). Lanka Rain Water Harvesting Forum & Community Water Supply and Sanitation Project.

Evans C. A., Coombes P. J., Dunstan R. H. and Harrison T. (2009). Extensive bacterial diversity indicates the potential operation of a dynamic micro-ecology within domestic rainwater storage systems. *Science of The Total Environment*, **407**, 5206–5215, https://doi.org/10.1016/j.scitotenv.2009.06.009

Heijnen H. (2013). Enhancing Economic Resilience in North Eastern Brazil by Harnessing Rain. Rainwater Harvesting Implementation Network. Rain Foundation, Amsterdam, The Netherlands. Retrieved from http://www.abcmac.org.br/files/downloads/heijnen_rainwater_harvesting_in_semiard_brazil_compressed.pdf

Heyworth J., Glonek G., Maynard E., Baghurst P. and Finlay-Jones J. (2006). Consumption of untreated tank rainwater and gastroenteritis among young children in South Australia. *International Journal of Epidemiology*, **35**, 1051–1058, https://doi.org/10.1093/ije/dyl105

IWMI. (2010). Managing Water for Rainfed Agriculture. International Water Management Institute. Retrieved from iwmi.cgiar.org/Publications/Water_Issue_Briefs/PDF/Water_Issue_Brief_10.pdf

JMP. (2017a). Progress on Drinking Water, Sanitation and Hygiene: 2017 Update and SDG Baselines. WHO/UNICEF Joint Monitoring Programme for Water Supply, Sanitation and Hygiene (JMP), Geneva. Retrieved from https://washdata.org/sites/default/files/documents/reports/2018-01/JMP-2017-report-final.pdf

JMP. (2017b). Safely Managed Drinking Water – Thematic Report on Drinking Water 2017. WHO/UNICEF Joint Monitoring Programme for Water Supply, Sanitation and Hygiene

(JMP). Retrieved from https://data.unicef.org/wp-content/uploads/2017/03/safely-managed-drinking-water-JMP-2017-1.pdf

JMP. (2019). Progress on Household Drinking Water, Sanitation and Hygiene 2000–2017. Special Focus on Inequalities. WHO/UNICEF Joint Monitoring Programme for Water Supply, Sanitation and Hygiene (JMP), New York.

Jones H. (2013). Mainstreaming disability and ageing in water, sanitation and hygiene programmes. Retrieved from https://wedc-knowledge.lboro.ac.uk/resources/learning/EI_WASH_ageing_disability_report.pdf

Melville-Shreeve, P., Ward, S. and Butler, D. (2016). Rainwater harvesting typologies for UK houses: A multi criteria analysis of system configurations. *Water (Switzerland)*, **8**(4). https://doi.org/10.3390/w8040129

Pathak N. and Heijnen H. (2007). Health and Hygiene Aspects of Rainwater for Drinking. Sustainable Development of Water Resources. Water Supply and Environmental Sanitation: Proceedings of the 32nd WEDC International Conference, Columbo, Sri Lanka.

Rautanen S. L., Laukka J., Pellinen S., Khadka S., White P. and Siddiqui F. (2015). HRBA & GESI STRATEGY & ACTION PLAN Human Rights-Based Approach and the Water and Sanitation Sector. DoLIDAR, Kathmandu, Nepal.

World Health Organization. (2017). Guidelines for Drinking Water Quality, Fourth edition incorporating the first addendum (4th ed). WHO, Geneva. Retrieved from http://www.who.int/water_sanitation_health/publications/gdwq4-with-add1-chapters/en/

Chapter 2

Harvesting the potential

B. Kniffen

American Rainwater Catchment System Association, Menard, Texas, United States

Keywords: harvest, rainwater, stormwater, sustainability

2.1 INTRODUCTION

The title of this chapter, 'Harvesting the Potential', relates to the possibilities that can be achieved by a unified organization such as the Mexico Rainwater Harvesting Association (Asociación Mexicana de Sistemas de Captación de Agua de Lluvia A. C., AMSCALL) through combining energy, talent, experience, and a desire to help one another, and is most important not only for this generation but even more so for the generations to come. This means not only harvesting the rain in containers, but also on the land, soil, and aquifers. The management of our watersheds affects the health of our cities and nations. Water is life and the ability to effectively manage our water resources is essential for personal, national and global sustainability. Without water there is no life. And on the flip side, too much water at one time impacts cities with floods and destruction.

In physics, for every action there is an equal and opposite reaction, but in nature, for every action there are many reactions. The impact of influencing natural processes has many rippling effects. Managing forests, savannahs, grasslands, and croplands is essential for health and sustainable processes on this planet. Landowners and controlling authorities of open space must be trained to care not

Editor(s): José Arturo Gleason Espíndola, César Augusto Casiano Flores, Raul Pacheco-Vega and Margarita Rosa Pacheco Montes
doi: 10.2166/9781789060584_0031

only for productivity but the long-term sustainability of the land, riparian areas, and the whole watershed. There is potential and opportunity to educate people that the small things they do in a united effort can have a huge impact on the environment in addition to policies made by governing authorities.

In urban and suburban areas, one rain garden and one rain barrel multiplied by many homes and businesses can have a huge effect on the watershed. Aquifers will be recharged.

With regard to the land, the health of a community is dependent on the health of the water and by reducing runoff one reduces pollution in our communities, streams and rivers. There is strength in numbers. Harvesting rainwater can be used to grow food, care for the native vegetation or rebuild forests. As a result of managing our watersheds one strengthens the health, wealth, and survival of those living in those communities. Growing some of our own food may mean food security in times of emergency, stress, or disasters. Collecting the rain to grow food makes us realize where food comes from and the importance of rainfall for the environment and our world's survival. In more arid locations harvesting rainwater may be a way for people to have better quality water to use and extend the life of a limited supply of water.

The knowledge of understanding how the rain can be captured, stored, filtered, and treated to make a potable drinking water supply can have a huge impact on families, their health, safety, and improve their quality of life.

Education may be the greatest need and opportunity the association can provide. Developing training programs to teach designers, installers, and decision makers as well as individual home and farm owners the processes, knowledge, construction techniques, and products available to capture, filter, and disinfect rainwater may have a huge impact on every person in the nation. Training individuals on processes needed to produce an effective rainwater harvesting design or installation, as well as teaching youth, volunteers, and groups, will spread the knowledge on the value of harvesting rain as well as help decision makers in make sound decisions.

2.2 HARVESTING RAINWATER FOR ENVIRONMENTAL STEWARDSHIP

The effects of climate change will impact us all. Most locations will face changes in rainfall patterns, rain amounts and intensity due to changing global conditions. Extended drought and dried up water supplies in arid and semi-arid locations affect survival itself. Growing forage for livestock and vegetables to feed a family, and having a product to sell all depend on water. Livestock and wildlife grazing have a significant impact on plant survival and overgrazing has a huge impact on vegetation survival and resilience such that when it does rain its ability to capture and infiltrate rainfall is affected.

Rangelands and forests must be managed properly. The Society for Range Management (SRM, https://rangelands.org) is a great resource to help one understand plant growth, effects of overgrazing and how to manage rangelands. SRM believes that the goals of maintaining the health and productivity of rangeland should supersede other management goals for long term ranch benefits. You can get eggs from a laying chicken for years, but you can only eat the chicken once. Rangeland is much the same, if it is not managed properly the more productive vegetation that sustains land health, water capture, and forage production is lost, often reverting the land to desert conditions that may never be restored in our lifetimes. Clearing forests also changes the direction of rainfall movement. Forests absorb water, recharge aquifers, cool, and protect the forest floor and the watershed. Quick dollars made from the destruction of forests may result in short term gains followed by negative impacts lasting generations.

The effects of climate change will impact us all. Many locations, croplands, and food production are greatly influenced by the amount of water captured in the soil. Rainfall leaving as runoff not only reduces the amount of water infiltrating and being stored in the soil for plant growth and food production, but runoff water contains not only the water itself but also topsoil, seeds, organic matter and nutrients which helps in food and crop production. Inorganic soil or soil without organic matter has minimal water storage ability. Organic matter is the sponge and keeper of water, nutrients, and biological life in the soil that supports healthy plant production. It must be maintained as is now seen in the USA in 'cover crop' promotion in the offseason to protect the soil, store, and hold water and add organic matter to the soil so crops grown for harvest are higher yielding. A nonprofit organization, Healing Hands International, (https://www.hhi.org/fighting-hunger) and other organizations teach how to utilize both organic matter from leaves and manure, and in combination with drip irrigation can help make food crops grow even in extreme situations.

Populations are growing in every major city in the world and most locations, this growth is on the fringes of our cities and on top of water infiltrating watersheds and recharge zones. Every meter of land impacted by urban sprawl has a compounding effect on everyone in the city and downstream. There are those visionary people and those willing to open their eyes can see the impact of this wound on the landscape that will leave a long-lasting scar. The more meters wounded, the greater the scar and the resulting impact.

> 'That land is a community is the basic concept of ecology, but that land is to be loved and respected is an extension of ethics'. – Aldo Leopold.

Our decision makers, as well as everyone who has raindrops fall on their place, has a responsibility. They and we have an ethical responsibility for future generations and the life of our cities for the next hundreds of years.

Aquifers all across Mexico and the world are being depleted. Almost 20 percent of the aquifers in Mexico are endangered or lost. That loss of stored aquifer water impacts water wells, springs, streams, and rivers and the quality and amount of water in our reservoirs and lakes. Water management must start at the top of the watershed. Clear streams, flowing springs, and healthy forests and lands depend on natures ability to manage every drop of water. Water must be allowed to infiltrate which requires vegetation in its healthiest state. The vegetation softens the impact of raindrops, leaves, stems of grasses and forbs and trunks of trees all slow water movement and filters the rainfall moving downhill. This allows rainfall time to infiltrate. The parts above ground are easy to recognize, but just as important are the roots below the ground. Roots die over time and plants replace them leaving channels for water movement. Organic matter is transported down into the soil by many creatures in a healthy environment, storing carbon, creating a sponge and pathways for water to move into this massive storage bank. Without care to the environment and vegetation, there is nothing to soften the impact of a raindrop, nothing to slow its movement across the watershed, and no help to allow infiltration and storage of the rain. Soil without organic matter becomes an impermeable barrier thus forcing water to runoff. This not only starves that land as it is not able to utilize that rainfall, but forces water away carrying topsoil, seeds, and nutrients with it. Every location must care for its raindrops for a healthy planet.

2.3 HARVESTING RAINWATER FOR STORMWATER REDUCTION

We all live in a watershed. Many cities are built near the bottom of a watershed and near a river, lake, or other reliable source of water. The increasing population around the world, expanding city population and expanding impervious area all result in increased stormwater, flooding, and disastrous situations on a larger and more frequent basis. As we seal over recharge zones, not only do we prevent water from infiltrating into the soil but we create human hazards of flooding, polluted streets, and rivers. Water which should infiltrate is forced to run off and impacting those below in the watershed. As the recharge zones and watersheds are sealed over with impervious roads, houses, and developments, and valuable forests, savannahs, and grasslands are abused or destroyed. In consequence, rainfall has no place to go but downhill with little to slow it down. The runoff creates energy as it moves downhill and rivers of concrete and pavement now replace filtering vegetation of a healthy ecosystem.

With urban development, aquifer recharge areas are sealed over which both decreases recharge of aquifers and increases the amount, speed and contaminant load as runoff moves down the watershed. City planning must involve protecting recharge zones. These areas are valuable assets that recharge underground supplies of freshwater and the best storage containers for our water needs, designing streets on a contour while developing infrastructure and greenspace which slows stormwater runoff and aids in water infiltration. There are many tools available to do this, but in residential areas, simple rain gardens are something every city can promote. American cities like Kansas City, Puget Sound, and San Francisco have promoted the construction of 10,000 rain garden projects in residential areas to reduce runoff. These were projects to help everyone understand that they are part of the problem and need to work to be part of the solution.

There are new innovations in water management that every city and individual can include to traditional stormwater management plans currently in place to reduce runoff. A term used for these practices is 'Low Impact Development' (LID). Each practice or multiple practices can be included in any location depending on the amount of impervious and pervious area available in the infiltration process, slope, local rainfall events, amounts and rain intensity. The key to their success depends on starting near the top of the watershed. That means everyone is in a watershed and everyone needs to be involved in its management. There is too much water even in desert locations to begin managing stormwater at the bottom of the watershed. Unfortunately, this is where many of our cities exist. It is too easy for city water planners to push runoff into the rivers, lakes or oceans as fast as they can while leaving aquifers dry. This causes municipalities to seek other sources of water outside their immediate area, robbing water from other areas to meet the needs of their growing base of customers which is not a viable long-term solution.

There are many processes often called 'Decentralized Urban Best Management Practices' (BMPs) which are designed to reduce the stormwater volume, reduce the stormwater flow rate, and reduce the pollutants in stormwater. There are tools and products available to tailor a process for most situations. Examples include curb cuts which allow water to exit our streets, permeable paving, water detention ponds that release water slowly, green roofs, rainwater harvesting, drop-in boxes, and many others. Applied at the point of stormwater generation and all along the downward slope of the watershed these tools will reduce or prevent flooding at the bottom of the watershed.

A rain garden is a tool that homeowners and businesses alike can install. It is a beautiful landscape feature consisting of a planted shallow depression that collects rainwater runoff from roofs, parking lots and other impervious surfaces. It is often the easiest homeowner storm-management tool. Kansas City and Marin County California have both implemented a program to construct 10,000 rain gardens in their area as a major tool to reduce stormwater runoff.

2.4 HARVESTING THE OPPORTUNITY TO INFLUENCE DECISION MAKERS

There is power in numbers. A united effort in rainwater harvesting from academia, manufacturers, designers, installers, researchers, users, and students will influence political affairs and policies. Policies that affect the watershed, environment, stormwater, and providing a sustainable source of water for every person in Mexico and the world. There need to be goals and a vision for a sustainable country, cities and rural areas. Education is powerful and adopting these tools will help move policy to care for Mexico's future.

As an organization grows and numbers of diverse individuals and companies increase, the impact the association will have will increase. Knowledge is powerful and knowing the problems, needs, and possible solutions will help influence those making critical decisions on a nation's future. Knowing issues locally and the magnitude of those issues can be a powerful tool to help decision makers make wise choices. Issues on water management are diverse. Some of those issues include the following.

Increased demand for a decreasing supply of water in both aquifer or surface water – In Texas, 40% of the springs once flowing have stopped which decreases base flow in the rivers, suppresses available water for humans, livestock, crops, and wildlife. In the USA lakes are filling from the bottom up with sediment which reduces the water storage capacity of lakes, the temperature of the lakes rise, evaporation increases and the value as a water supply or stormwater retention resource decreases.

Escalating environmental and economic costs in the face of an aging infrastructure, leaking and broken pipes, environmental repair or protection of sensitive areas – As cities grow, the increased demand for water requires infrastructure to deliver clean water while removing wastewater. Most cities are aging as is the infrastructure. Most cities were not designed for the growth they have seen in the last 100 years. The aging pipes, streets, buildings, and ignored natural waterways cause untold damage while leaking pipes increase waste and contamination in the cities. Replacing and repairing these issues is expensive and often there is not enough money to rebuild systems only exacerbating the problem.

Climate change, drought, flooding – What does the future hold? Drought, floods, and seasonal changes in rainfall? Combining this with a growing population makes current methods of meeting water needs, while protecting water sources, impossible. There must be a paradigm shift in managing rainfall. Rainfall is the free gift of nature and to be cherished.

2.5 HARVESTING RAINWATER FOR LIFE

The association can also help harvest the potential and opportunity to capture rain for people without a good water source. Many places in the world in the arid or semi-arid regions, as well as regions with short monsoon seasons followed by several months without any significant rainfall, may run out of water and sources dry up or may be or become contaminated as the supply dwindles and/or utilized by livestock and wildlife who also uses these supply sources. By harvesting the rain in containers when it does rain and either using that harvested rain first (or last) can extend the life of a limited supply of water. Providing a reliable source of good quality water is one of the most important gifts an association can give people. The collected rain could be used as a family's non-potable source where a limited but quality supply exists like in a water well or borehole. Harvested rain can be filtered and treated so it is a safe source of potable water available for families. There are several methods available not addressed in this chapter which may be applied to reduce or eliminate the risk of water carrying contaminants which may make people and animals sick.

Food production requires water. Capturing and storing water in the soil is the cheapest and most important source of water to grow crops, native grasslands, savannahs, and forests while recharging aquifers. But harvested in containers there is little evaporation or loss and every drop can be directed for its intended need. This is particularly valuable in growing vegetables with drip irrigation, hydroponics or hand watering. Many of these techniques may be found in the efforts of NGOs such as the Rain Foundation (http://www.rainfoundation.org/).

Desertification of our semi-arid areas where water once was and areas with seasonal extremes of a monsoon season followed by many months without rainfall are ideal candidates for harvesting rainfall when it comes. Rooftops that only contribute to runoff can be converted to rain-harvesting mini watersheds for collection for later use.

Many places in the world in the arid or semi-arid regions, as well as regions with alternating monsoon seasons followed by several months without any significant rainfall, may run out of water and sources dry up or water may be contaminated or becomes contaminated as the supply dwindles and/or utilized by livestock and wildlife who also uses these supply source. By harvesting the rain in containers when it arrives and either using that harvested rain first (or last) can extend the life of a limited supply of water. Supplying clean water to a family living with contaminated water as their only source is one of the most important gifts an organization can provide. We say 'water is life' but for many people, the only source of water is contaminated with all kinds of pathogens. Untreated water can cause sickness, unproductivity, and too often death to the youngest and oldest in a population. Harvested rainwater may be the best water source available to many in the world.

In Texas, rainwater is becoming the 'Gold Standard' for water in our homes. Without salts, minerals, and most other impurities, rainwater is ideal for home devices and our bodies. It leaves water heaters, washing machines, and dish-washers. The soft water is ideal for bathing and washing dishes. Harvesting rainwater for in-home use has increased, the rainwater is ideal for house plants and all daily uses.

2.6 HARVEST THE POTENTIAL TO PROVIDE EDUCATION

Education may be the greatest need and opportunity an association can accomplish. Developing training programs to teach designers, installers, and decision-makers as well as individual home and farm owners the processes, knowledge, construction techniques and products available to capture, filter, and disinfect rainwater may have a huge impact on every person in the nation. Training individuals on processes needed for effective rain harvesting design or installation, as well as teaching youth, volunteers and groups, will spread the knowledge on the value of harvesting the rain as well as help decision makers in making sound decisions.

2.6.1 The American Rainwater Catchment Systems Association (ARCSA)

The American Rainwater Catchment Systems Association (ARCSA) has the stated mission 'to promote sustainable rainwater harvesting practices to help solve potable, non-potable, stormwater and energy challenges throughout the world'. Efforts to accomplish this are achieved through an annual National Conference, training programs, regional representatives and a website with a resource directory, project gallery, and much more. ARCSA had its beginning around 1985 in Austin, Texas but struggled to find its purpose until 2003 when it held its first national conference.

That one-day conference of presentations, vendor displays, and an open planning meeting led to its second conference in Seattle in 2005. Speakers, vendors, a second-day tour of existing systems and discussion on how to expand the association's efforts led to the formation of regional representatives, a slate of officers and officer planning meetings.

The third national conference in 2007 was held in Hawaii and it was there that the organization found direction, meaning, and manpower. The educational program was born and training sessions began. The first training was a full-day training program but it soon expanded into a two-day program with two primary instructors, Billy Kniffen, with Texas A&M University and Tim Pope, an ARCSA past president, and installer in the Seattle area. In 2010 a second two-day program was added and called the 'Design and Construction Workshop' and was a hands-on section to implement the knowledge gained from the first two-day lecture program. Students are divided into groups or small companies of 2–4 people, given a project and have to design a rainwater system to meet the needs of the project including managing the overflow water. This becomes a peer training program as students teach each how to do the work, solidifying the previous classes training. They must calculate sizing requirements, potential rainfall, make sketches and present their proposal in a mock bidding war, presenting their bid and sharing their design. Students then critique each proposal. Students then construct two smaller systems to get familiar with tools, supplies, products used in constructing systems with the second including a pump which is installed and tested. A third project is a class effort to install a much larger more complex system and students then develop a start-up, operation, and maintenance manual for the operator/owner of the installed system.

The third phase of the program began in 2011 with the addition of a one-day show to a training program on how to inspect an existing rainwater system. Those who wanted to be recognized as a 'Rainwater Inspection Specialist' must have gone through the first four days of training and passed a written test and provided to ARCSA information on five systems they have been involved with, reviewed and approved. The inspection program begins with a review of system requirements, calculations, review of a checklist, evaluation protocol, and checklist on parts inspected. There is also a written test for this recognition.

The first training manual (Kniffen et al., 2012) was developed through the Biological and Agricultural Engineering Department at Texas A&M in 2009–10 and has been translated into Spanish by Professor Arturo Gleason in 2014. It has been used in training programs all across the USA until 2018 when ARCSA developed their own training manual (Audrey, 2015) through the efforts of volunteer members writing sections in their field of expertise. The educational program has grown into three major sections:

(1) 'Accredited Professional' rainwater harvesting course is designed for industry professionals and those individuals desiring to pursue a career in rainwater management. The completion of this course is one of the requirements for attaining recognition as an ARCSA Accredited Professional. An alternative to traditional water management systems.
(2) The 'Master' level in rainwater harvesting is ARCSA's highest level of recognition. 42 hours of additional training is required and many of those can be gained through ARCSA webinars which are also open to anyone seeking more information on specific topics. The Design and Construction (D/C) Workshop and Inspection Specialist courses are also required.
(3) ARCSA 'Inspection Specialist' program for ARCSA APs (others may attend the class) is designed to develop a formal inspection of systems to assist the industry and individuals. Must also have attended the D/C Workshop to be an IS.

A Design/Construction Workshop is set out to give hands-on experience in both design and construction of basic systems and procedures for developing operation manuals and troubleshooting problems. All these courses can be found on the ARCSA website (www.arcsa.org). The Accredited Professional is also an online course.

2.7 CONCLUSION

Sustainable high-quality water supply needs to be provided to every person in the world. Certainly aquifer, river, and lake water will supply the needs of most people. The cheapest and best option should be used, but not at the cost of destroying aquifers, polluting rivers and lakes, and destroying the health and well-being of future generations. Every city must look long term and make sound decisions for our future. There are options, and these options should include rainwater harvesting, which is a viable potable water source that can solve many other environmental challenges. Captured rainwater is also a valuable necessity in sustainable agriculture which will help replenish aquifers rather than deplete them, extending food production for everyone, whether on rooftop gardens or larger commercial operations. The quality of rainwater has been declared the 'Gold Standard' for irrigation.

Pope Francis urged world leaders at the United Nations to act as compassionate custodians on issues including Europe's migrant crisis, world conflict and climate change. 'We human beings are part of the environment', Francis said. 'We live in communion with it, since the environment itself entails ethical limits which human activity must acknowledge and respect … Any harm done to the environment, therefore, is harm done to humanity'. On the environment, Francis said 'Christians believe with other religions that man is supposed to take care of nature. He is not authorized to abuse it and much less is he authorized to destroy it'.

May God bless the efforts of so many whose hearts reach out to care for making Mexico and the world a sustainable planet that has a compassion not only for every citizen today, but children of generations to come. Sustainable decisions are not easy. The united efforts of people learning from others and applying this knowledge in a united effort of the organizations like the Mexico Rainwater Association (AMSCALL) will truly ensure the future looks bright for Mexico.

REFERENCES

Audrey A. (2015). Rainwater Harvesting Manual. American Rainwater Catchment Systems Association, Tempe, AZ, USA.

Kniffen B., Clayton B., Kingman D. and Jaber F. (2012). Rainwater Harvesting: System Planning. Texas AgriLife Extension Service, B6240, USA.

Chapter 3

Transforming tradition of rainwater harvesting in Sri Lanka

T. N. Ariyananda

Lanka Rain Water Harvesting Forum, Sri Lanka

Keywords: tank cascade system, rainwater harvesting, rainwater policy, disasters, climate change, health

3.1 INTRODUCTION

Sri Lanka is a tropical island of 65,000 sq.km located in the Indian Ocean and rain is the primary source of water. Sri Lanka has an average annual rainfall of 1800 mm which ranges from 900 mm in the dry zone to 5000 mm in the wet zone. The rainfall is bi-monsoonal and varies both seasonally and spatially. According to rainfall, the country is divided into three zones viz. dry zone, intermediate zone and wet zone. The dry zone covers nearly two thirds of the total land extent of the country, where the elevation is less than 300 m and receives an annual rainfall less than 1750 mm (Punyawardane, 2008) (Figure 3.1). Annual evaporation in the dry zone ranges between 1700 and 1900 mm which results in a soil moisture deficit during dry periods (Panabokke *et al.*, 2002). Dry zone is characterized by a long dry spell (nearly eight months in some cases from Feb. to Oct.) and a short rainy season (nearly four months from Sep. to Jan.), during which 80% of the annual rainfall is received. The dry season in the dry zone is plagued by chronic and recurrent droughts and desiccating winds, while seasonal flooding dominates the rainy season. The water problem in the dry zone is further aggravated by the low water retention capacity of the unique soil group (reddish brown earth) which dominates most parts of the dry zone (Mapa *et al.*, 2010).

Editor(s): José Arturo Gleason Espíndola, César Augusto Casiano Flores, Raul Pacheco-Vega and Margarita Rosa Pacheco Montes
doi: 10.2166/9781789060584_0041

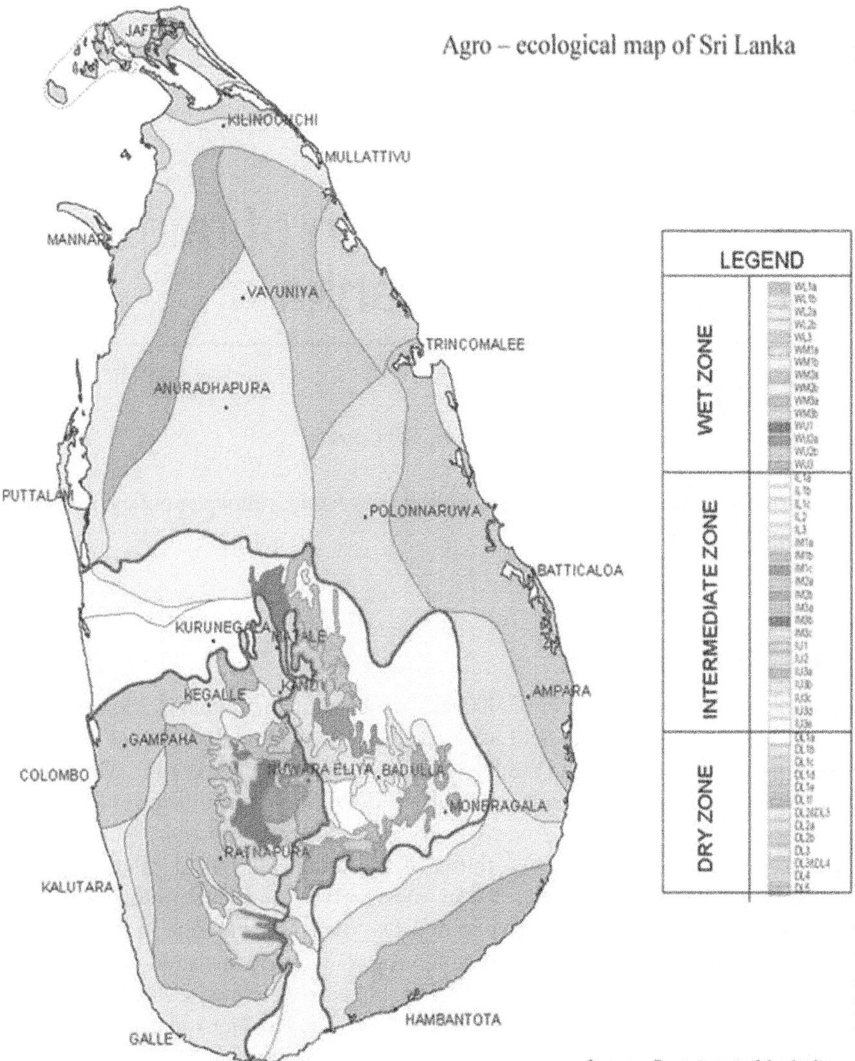

Figure 3.1 Agro-ecological map of Sri Lanka. (*Source*: Punyawardena B.V R. Bandara T.M. Munasinghe M.A.K. Banda. N.J (2003)).

Temporal and spatial disparities in annual rainfall led the ancient rulers of Sri Lanka to develop the tank cascading system (Figure 3.2). The main principle behind the Tank Cascade Systems (TCSs) is recycling and reuse of water through a network of small to large scale tanks. Brohier (1935) has also reported village

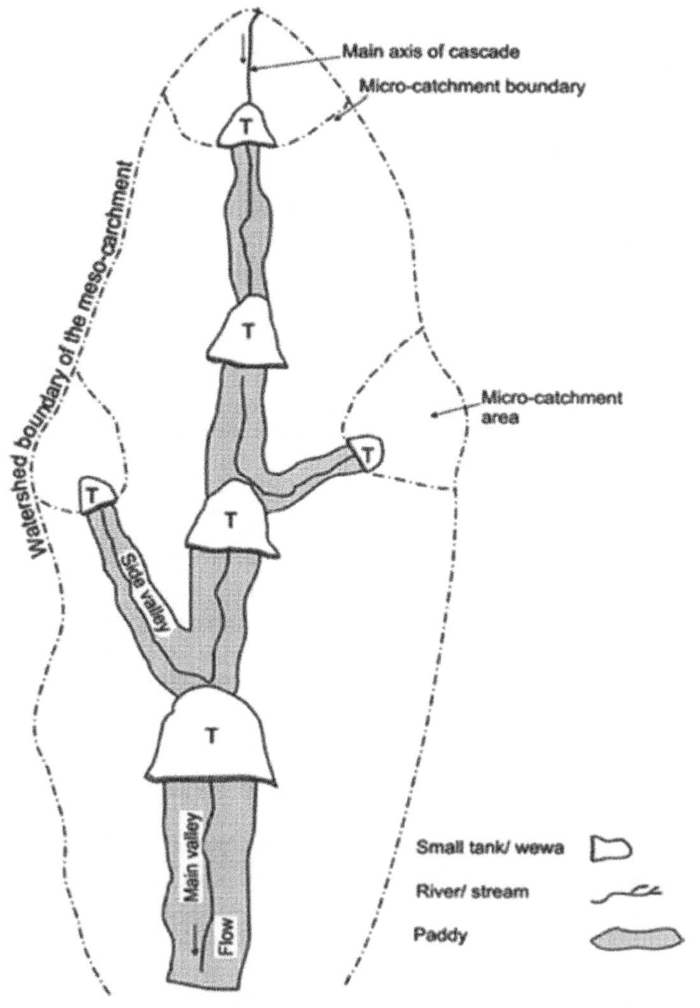

Figure 3.2 Schematic representation of a small Tank Cascade System. (Adapted from Panabokke *et al.*, 2002).

tanks and their association with ancient irrigation works in Sri Lanka. Tank cascading system controlled the seasonal flooding and drought in the dry zone and also ensure the continuous cultivation of paddy during both rainy and dry seasons and provide a seasonal water supply for domestic and livestock consumption.

The first great reservoir of the world ever constructed was Panduwewa or Panda Wewa (1,360 acres) of Sri Lanka built by King Dappula II (807–812 AD) according to Henry Parker, a British colonial irrigation engineer (1873–1904) in British

Ceylon (1815–1948). There are around 30,000 such tanks or wewa (Figure 3.3) constructed in the dry zone from which the majority was built during 3rd–12th century BC. Our ancestors conserved water not only for irrigation and domestic purposes, but also for conservation of the environment, retention of soil moisture and maintenance of the water table. The famous quote of the Sri Lanka King, Parakramabahu the Great (1153–86 AD) that 'Let's not allow a single drop of water to flow into the sea without being used for the benefit of mankind' (Arumugam, 1969 quoted from Mahawansa), is an indication of the commitment of the ancient rulers and the people to conserve water and minimize the runoff. Further, use of rainwater by kings for landscaping and aesthetics is seen at the 5th Century rock fortress of Sigiriya (Figure 3.4) comprising man-made rainwater reservoir feeding the pools, ponds and fountains by means of an underground network of conduits.

There are many traditional domestic rainwater systems also reported and still in operation in Sri Lanka (Ariyabandu R D.S, 1998). They are: (1) tree trunk collection where rainwater is collected for domestic use from tree trunks using banana or coconut leafs specially in the hilly areas in the wet zone (Figure 3.5); (2) collection into pots and buckets where rainwater is collected during the rainy period by using the roof (Figure 3.6) or temporary collection surface such as cloth or sari tied between four poles and a pot kept in the middle; and (3) brick tank collection where an open square brick tank placed on the side of the dry zone houses collects water from the roof for use using temporary gutters.

Over the last 25 years, traditional rainwater harvesting has been revived to address the acute water shortage experience due to temporal and spatial variation in rain fall and climate change disasters and many research studies were conducted to improve the technology. In 19995 Community Water Supply and Sanitation project initiated by government of Sri Lanka with World Bank funds

Figure 3.3 Tank or *wewa* in North central Province. (*Source*: LRWHF, s.f).

Figure 3.4 Sigiriya Rock Fortress. (*Source*: Tennakoon, Amila. *Sigirita Rock Fortress*. 2012. https://flic.kr/p/cVaSBo.

Figure 3.5 Rainwater collected from tree trunks. (*Source*: LRWHF, s.f).

Figure 3.6 Rainwater collected from roof top into containers. (*Source*: LRWHF, s.f).

introduced rainwater harvesting as a water supply option in two wet zone districts Badulla and Matara. Since then, government and non-government organisations throughout the country have promoted the technology. There has been a significant increase in the use of roof water harvesting in Sri Lanka, which has proved to be a boon to rural people, particularly for domestic water supplies in water scarce situations. As a result, presently there are estimated 42,000 domestic rainwater systems which has brought much relief to households during times of drought, floods, tsunami, chronic kidney disease of unknown etiology and resettlement for many people living in rural areas.

3.2 INSTITUTIONAL UPTAKE OF RWH

In 1992, the government of Sri Lanka initiated Community Water Supply and Sanitation Project (CWSSP) with assistance from the World Bank with an objective of providing affordable safe drinking water and sanitation facilities to rural community in three districts in the wet zone and hilly area of the country: Badulla, Matara and Ratnapura districts. During the implementation of CWSSP, it was clearly identified that several villages in Badulla district could not be served with available conventional water supply technology. For these uphill settlements, rainwater harvesting was thought to be a possible option to provide water with minimum cost.

The feasibility study on rainwater harvesting was conducted in 1995 (Hapugoda, 1995). The study, after much experimenting was able to design and construct two types of tanks that could be built within the financial limits of the project (in

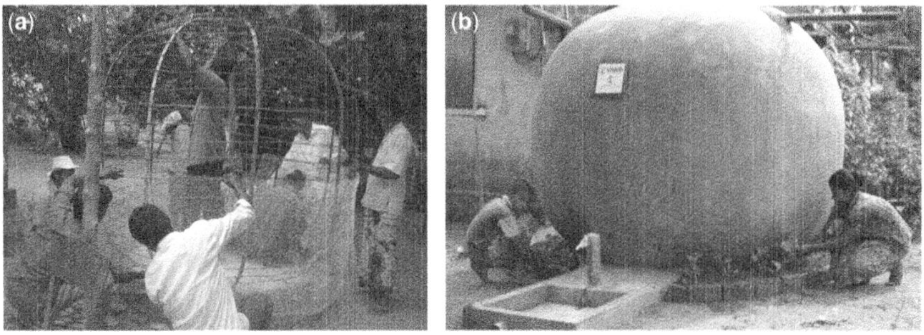

Figure 3.7 (a) Constructing a ferrocement pumpkin tank. (*Source*: LRWHF, s.f). (b) Finished ferrocement pumpkin tank. (*Source*: LRWHF, s.f).

1995, US$100). One tank made of ferrocement, has the shape of a pumpkin (Figure 3.7a, b) and is unique to the project. The second tank is built under ground and is an adaptation of the Chinese style biogas pit, using a simple, low cost technique to do the dome roof.

The special feature in the ferrocement pumpkin tank is the skeleton mould made of eight or ten 'T' iron legs and a 6 mm re-enforcement bar. At the end of construction both the legs and the 6 mm bar are removed leaving a pumpkin-shaped tank (so far tanks can be made up to 10 cubic meters) made of 40 mm thin wall reinforced only with two layers of 12.5 mm chicken mesh.

Initially there was a great deal of protest to the project due to misconceptions of rainwater harvesting. A petition was also submitted to the Minister in charge of water supply. However, after the successful demonstration of one pilot project in a village called Dematawelihina in Badulla district, there were 146 applications from the village. Since then there has been 5350 rainwater tanks constructed in the Badulla district.

Since the initiation of the CWSSP, a number of organizations and institutions have adopted rainwater harvesting as a means of supplying water to water scarce households in both the wet and dry zones. Some of the noteworthy contributions in rainwater harvesting for domestic use have been made by: the Southern Development Authority; the Dry Zone Development Project funded by IFAD (International Fund for Agricultural Development); National Water Supply and Drainage Board (Chronic Kidney Disease Project, 3rd and 4th ADB Water and Sanitation projects); and the 2nd CWSSP project.

3.3 LANKA RAINWATER HARVESTING FORUM

The CWSSP project initiated the emergence of the Lanka Rain Water Harvesting Forum (LRWHF), which is an NGO actively engaged in promoting rainwater harvesting in the country. In early 1996, a small group of persons interested in rainwater harvesting, who aspired to organise a network to pool and share the experience of the country, came together to form the Lanka Rain Water Harvesting Forum (LRWHF). The group consisting of members from government, non-government, research and private sector organisations placed advertisements in the local papers (local languages) to gather interest and experience in the country. With the encouraging response received from the public the group considered its role and potential in promoting the use of rainwater harvesting and established itself with the following objectives:

- Identify existing rainwater harvesting practices in the country
- Develop further techniques for collection of rainwater
- Promote the application of rainwater for domestic purposes through information, communication, and awareness raising
- Initiate further studies to make recommendations for maintaining good rainwater harvesting practices.

During the last 23 years LRWHF has actively promoted the concept of rainwater harvesting in all districts through demonstration projects, awareness programs, training, research and development, and networking. A large number of masons (500) and professionals/officials (around 400) were trained in design, planning and construction of rainwater harvesting systems in many districts. Awareness programs conducted by the Forum in the villages, media as well as in the form of seminars created much public and official interest in rainwater harvesting. Forum also conducted research studies to improve on the present technology and to study water quality and water user patterns of the rainwater-harvesting units. Information generated by these studies is disseminated through seminars and web site.

3.4 RAINWATER POLICY AND REGULATIONS IN SRI LANKA

As a result of influencing and lobbing carried out by LRWHF, in 2005 the Hon. Minister for Water Supply and Urban Development appointed a committee to formulate a national policy on rainwater harvesting and strategies. LRWHF members played an active role in formulating the national strategy and policy on rainwater harvesting. Public as well as government ministries and other relevant authorities were invited to comment on the draft policy and make suggestions to improve it. In June 2005, the government of Sri Lanka passed the world's first national policy on rainwater harvesting (National Rain Water Harvesting Policy and Strategy, 2015). The policy objective is aimed at encouraging communities to control water near its source by harvesting rainwater. This resulted in: minimizing the use of treated water for secondary purposes; reduction of flooding; improving soil conservation and groundwater recharge; providing water for domestic use with adequate treatment; agricultural benefits and reduced energy consumption.

This policy document, states the required legislative changes needed to amend the Urban Development Authority (UDA) and Road Development Authority (RDA) by laws on drainage and National Water Supply & Drainage Board (NWSDB) by-laws to incorporate harvesting rainwater as a source of domestic water. The regulation has been gazetted on the 17th April 2009, which makes rainwater harvesting mandatory in certain categories of new buildings (and old building after 3 years) in areas under municipal and urban council jurisdiction (Urban Development Authority Act, 2009).

3.5 PROVIDING WATER AND FOOD SECURITY FOR COMMUNITIES AT TIMES OF DISASTER

During the tsunami of 26th December 2004 an estimated ten pipe-bourn water schemes and 50,000 house connections were damaged. In addition, it is estimated that nearly 40,000 wells (60%) were contaminated bacteriologically and with salinity (Saltori & Giusti, 2006). Therefore, the people living in these areas faced serious problems in fetching quality water for domestic use. Especially since the bowsers supply of water by the donor agencies and government was also soon diminished.

LRWHF implemented a project to introduce rainwater harvesting systems (RWHS) in households, schools and institutes in three of the tsunami affected districts in the southern and eastern coastal areas in Sri Lanka. Around 4000 rainwater-harvesting units have been built for the tsunami reconstructed houses in two districts in the south and one district in the east. RWH systems significantly addressed the shortage of domestic water at household levels by providing easy access to clean drinking water, less time spent on collecting water, skilled enhancement in the village, less reliance on external water providers, more water security at household level and better sanitation practice due to more water being available. The system also supported the improvement and maintainance of a home garden of the newly resettled community, thus providing livelihood and improving food security (Figure 3.8).

Government policy and support has also encouraged the uptake of rainwater harvesting in the tsunami reconstruction process. Ministry of Urban Development and Water Supply incorporated rainwater harvesting in the design of tsunami reconstructed houses, which was an essential requirement in a standard tsunami house.

In 2009, after the end of 30 years of civil war in Sri Lanka, the communities resettled either in their own homes or in newly built homes in the Northern Province. Ground water is the main source of potable water in most of this area, which is been over exploited or polluted due to excessive use of fertilizers. In 2011, LRWHF with the support of USAID introduced a program to install RWH systems in 850 households, 31 schools and 5 hospitals. Home garden development and plant nursery centers were also introduced during this project.

Figure 3.8 Home gardening with RWH. (*Source*: LRWHF, s.f).

The program has helped communities in the former northern conflict zone return to normalcy as quickly as possible through restoration of water sanitation and hygiene, food security and livelihoods.

3.6 CLIMATE RISK RESILIENCE

During the last few years, the frequency of climate disaster occurrence in the country has caused water stresses to people in almost all districts due to drought and floods. Heavy rainfall, floods and long droughts are frequent occurrences and have increased significantly over the past ten years, leading to Sri Lanka being positioned second in the recent ranking based on vulnerability to climate change (Global Climate Risk Index, 2019).

Lanka Rain Water Harvesting Forum in partnership with USAID constructed rainwater harvesting systems in households (360), schools (45) and hospitals (10) in the drought and flood prone areas in the Northern Province and Uva Province of Sri Lanka. It has been reported that during the drought and floods in recent years these households and institutes installed with rainwater harvesting systems were able to cope better. During the droughts experienced over 2017–2018, households which had rainwater collected and stored were better off since they had readily available water at home. In 2018, during floods in Kilinochchi district when the surface water sources and well water were contaminated, the only clean water source that was available to the community was the rainwater harvesting tank. Through this project disaster risk reduction measures and tools have been introduced to local communities. More than 10,000 persons are provided with access to safe drinking water through RWH.

Figure 3.9 RWH systems constructed in schools. (*Source*: LRWHF, s.f).

RWHS built in schools, which function as flood shelters, provide water to the people until government aid can reach them (Figure 3.9). Around 100 RWHS were constructed by LRWHF and maintaining groups are set up with students, teachers and parents as a sustainability measure. A manual for operation and maintenance of RWHS in schools was published by LRWHF with collaboration of PLAN SL and National Water Supply & Drainage Board to provide practical solutions and guidance for school teachers, parents, caretakers and students for operation and maintain RWHS to ensure long term sustainability.

Lanka Rain Water Harvesting Forum, recognizing the importance of knowing about the day to day weather data by the community in order to prepare for extreme events, has made available weather stations in 12 remote schools in eight districts of Sri Lanka. Low cost solar-powered mobile weather stations have been developed by International Water Management Institute (IWMI) in collaboration with Moratuwa University which measure weather data such as temperature, rainfall, wind direction, wind speed, pressure and humidity. The weather stations are established in schools to build awareness among school children about weather and weather changes. LRWHF has trained school children on how to operate and maintain these weather stations. The weather data is shared with all students and teachers in the school through the display boards provided. The data, which play an important role in the daily lives and livelihoods of rural community members previously not accessible by the community, is now provided at their fingertips. This information is communicated to the community through SMS or mobile application to be better prepared for these crisis situations. The data collected is also fed into a wunderground web site (https://www.wunderground.com/) which also gives ten-day weather prediction. This data can then be used for decision making in regular water resource management and other climate-related application including helping to avoid crises situation, such as droughts and floods. The schools installed with mobile weather station are able to better manage the water in the rainwater harvesting tank by knowing about the rainfall patterns in the area. This will help them to conserve during the dry periods and to use up tank water before the rainy season.

3.7 IMPROVING HEALTH

An alarming and new form of chronic kidney disease of an unknown origin (CKDu) has been reported in several areas of Sri Lanka. The total number of affected individuals is estimated to be around 70,000, and reports a death toll of 20,000 so far (Edirisinghea *et al.*, 2017). Despite many studies conducted, the causal factor of this disease is still uncertain (thus 'u' in CKDu). There is widespread consensus among scientists that improving the quality of drinking water in these areas may prevent, or retard the progress of this kidney disease.

Survey of RWH system constructed in households with CKDu patients reports that conditions of patients at early stage of CKDu either remains the same or has reversed by drinking rainwater (LRWHF, 2016). However, it is yet to be clinically proven. High demand for RWH from people in these areas is evidence of the anxiety to obtain clean drinking water.

A study conducted in North Central Province where most CKDu patients have been recorded, indicated that harvested rainwater quality was better (bacteriologically and chemically) compared with well and reservoir water tested (Aioma, 2018). A recent survey recorded that 52.8% of households with CKDu patients drink rainwater (UNDP, 2018).

3.8 RUNOFF COLLECTION FOR CROP GROWTH

The rural sector in Sri Lanka constitutes around 80% of the population and most of those in this sector depend on rainfall-based sources of income, such as agriculture, livestock production and inland fisheries. Freshwater availability is a key limiting factor in food production and improvement of livelihood.

Lack of a dependable water supply is a major limiting factor in attempts to develop the rural sector. From the total rainfall, on average around 60% of rainwater is lost in the form of surface runoff and conserving this water will promote crop growth in areas where water is limited.

In some parts of the dry zone, small ponds called 'Pathahas' have been used traditionally to collect and store rainwater (Figure 3.10). Such a water collecting system on a farm has enabled farmers to cultivate crops during the dry seasons. These ponds vary from 300 to 500 m^3. Usually ponds are constructed in a valley, at the bottom of the cultivated land to facilitate the gravitational flow of runoff water via contour drains towards the ponds for collection. Harvested water is lifted out of the pond manually by use of a bucket or using pedal and kerosene pumps.

Figure 3.10 Garden pond or 'pathahas'. (*Source*: LRWHF, s.f).

Studies have shown that collection of runoff water can be used for agriculture as well as to improve the ground water levels both qualitatively and quantitatively. A study was carried out by LRWHF in Kurundamkulama (a village in Mihintale in Anuradhapura District to harvest/collect runoff rainwater in a 5 m^3 underground tank enabling the farmers to cultivate a crop during Yala (lesser rain season). As a result, the incomes of the families in the study area increased substantially (Weerasinghe *et al.*, 2005). Collection of runoff rainwater in this manner not only conserves water but also reduces soil erosion and degradation of the land.

Another study conducted by LRWHF in Nikaweratiya (de Silva, 2005) has shown that collecting rainwater in ponds or pathahas, as in the ancient systems, elevates the ground water level by allowing the water to percolate into the ground, thus, increasing the quantity of water available for both domestic and agricultural use even during the dry season.

3.9 CONCLUSION

Rainwater is a valuable resource, which should be exploited in the most efficient way. In order to achieve SGD6 of universal access to safely managed water by 2030, rainwater harvesting is a must in Sri Lanka and many other South Asian and African countries. Rainwater harvesting can also facilitate achieving seven more of the 17 goals, namely: SDG1 (No Poverty), SDG2 (Zero Hunger), SDG 3 (Good Health and Well-being), SDG5 (Gender Equality), SDG 8 (Decent Work and Economic Growth), SDG11 (Sustainable Cities and Communities) and SDG13 (Climate Action). Governments should promote and implement rainwater harvesting policy to encourage rainwater harvesting for human consumption and domestic use, to protect the people's health and livelihood, ensure good utilization and conservation of water resources, adapt for climate change disasters and facilitate guidance and capacity building.

REFERENCES

Aioma G. V. N. (2018). Rain Water as a Source of Safe Water in Northern Central Province. International Conference on Water Security through Rain Water Harvesting, LRWHF, Colombo, Sri Lanka.

Ariyabandu, R. de S. (1998). Study of existing rainwater harvesting technology prepared for the Lanka Rainwater Harvesting Forum, unpublished.

Arumugam S. (1969). Water Resources of Ceylon. Water Resources Board, Sri Lanka.

Brohier R. L. (1935). Ancient Irrigation Work of Ceylon (Sri Lanka), 1st ed. (Translated by Piyasena L.) Mahaweli Center, Mahaweli Authority of Sri Lanka, Colombo.

de Silva S. (2005). Impact of Artificial Recharging of Groundwater with Rainwater: a Case Study in Kotewehera of Sri Lanka, Lanka Rain Water Harvesting Forum, Sri Lanka.

Edirisinghea E. A. N. V., Manthrithilake H. and Pitawalac H. M. T. G. A. (2017). Geochemical and isotopic evidences from groundwater and surface water for understanding of natural contamination in chronic kidney disease of unknown etiology (CKDu) endemic zones in Sri Lanka. *Isotopes in Environmental and Health Studies* 2017 https://doi.org/10.1080/10256016.2017.1377704 (accessed 15th July 2019).

Global Climate Risk Index. (2019). David Eckstein, Marie-Lena Hutfils and Maik Winges, Germanwatch e.V, www.germanwatch.org/en/cri, www.germanwatch.org/en/cri (accessed 15th July 2019).

Hapugoda K. D. (1995). Action Research Study on Rain Water Harvesting. Project Report 2. CWSSP, Ministry of Housing and Construction. Sri Lanka.

LRWHF. (2016). Survey of rainwater harvesting households with CKDu patients in Mahakachchakody, Vavuniya District 2016. Lanka Rain Water Harvesting Forum, Sri Lanka.

Mapa R. B., Somasiri S. and Dassanayake A. R. (2010). Soils of the Dry Zone of Sri Lanka; Morphology Characterization and Classification. Soil Science Society of Sri Lanka, Peradeniya, Sri Lanka.

National Rain Water Harvesting Policy and Strategy. (2015). Ministry of Urban Development and Water Supply, Sri Lanka, www.lankarainwater.org/rwhsl/nrwpolicy-2005.pdf (accessed 15th July 2019).

Panabokke C. R., Sakthivadivel R. and Weerasinghe A. D. (2002). Evolution, Present Status and Issues Concerning Small Tank Systems in Sri Lanka. International Water Management Institute, Colombo, Sri Lanka.

Punyawardena B. V. R., Bandara T. M., Munasinghe M. A. K. and Banda N. J. (2003). Agro Ecological Regions of Sri Lanka. Department of Agriculture, Sri Lanka.

Punyawardane B. V. R. (2008). Rainfall in Sri Lanka and Agroecological Zones (Sinhala Version). Department of Agriculture Press, Peradeniya, Sri Lanka.

Saltori R. and Giusti A. (2006). Challenges of tsunami and conflict affected rural water supply in Sri Lanka. 32nd WEDC International Conference, Colombo, Sri Lanka, 2006.

UNDP. (2018). Survey on the Perception of Users of Rain Water Harvesting Systems in the Seven Project Districts, Climate Resilient Integrated Water Management Project, Lanka Rain Water Harvesting Forum, Sri Lanka.

Urban Development Authority Act, Extra Ordinary Gazette No. 1597/08 on 17-04-2009 of Social Republic of Sri Lanka, http://lankarainwater.org/wp/?page_id=72 (accessed 15th July 2019).

Weerasinghe P. A., Ariyananda T. N. and Weeraratna C. S. 2005 Rainwater Harvesting for Home Gardens in Dry Zone of Sri Lanka. Proceeding of XII IRSCA Conference, New Delhi, India.

Chapter 4

Rainwater catchment on Hawai'i Island

Patricia S. Macomber
Keeau, Hawaii USA
macomber@hawaii.edu
Editing assistance by John Hammerstrom

Keywords: catchment, Hawaii, rain, rainwater, water, sustainability

4.1 INTRODUCTION

Hawai'i Island is one of the few places in the United States where rainwater catchment is common and necessary. Much of Hawai'i Island is rural and lacks public utilities including water. Rainwater catchment systems are the most customary alternative. The Rainwater Catchment Program of the University of Hawai'i was instrumental in enabling catchment owners to understand their water supply risks and how to properly care for their systems in order to improve water quality and quantity. This chapter introduces readers to the unique rainwater collection history and usage of Hawai'i, and describes the rainwater catchment program of the University of Hawai'i, how it developed and serviced the residents of the islands. It also discusses challenges such as lack of system standards and water quality threats. The rainwater catchment program grew and even touched international arenas, addressing issues well beyond its humble beginnings.

Editor(s): José Arturo Gleason Espíndola, César Augusto Casiano Flores, Raul Pacheco-Vega
and Margarita Rosa Pacheco Montes
doi: 10.2166/9781789060584_0055

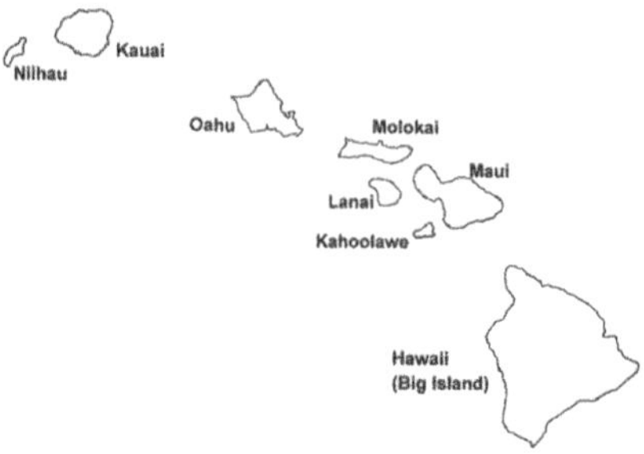

Figure 4.1 Major Hawaiian Islands. (*Source*: HFD, 2008).

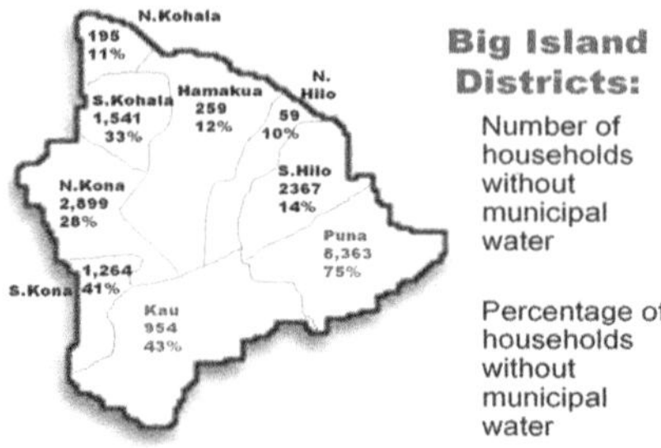

Figure 4.2 Municipal water availability. (*Source*: HFD, 2008).

4.1.1 Introduction to Hawai'i

The USA state of Hawai'i has 8 major islands (Figure 4.1). The islands are volcanic in nature, formed as lava erupted from a 'hot spot' near the middle of the Pacific Tectonic Plate. The largest of the 8 major islands shares its name with the state, so to avoid confusion Hawaii Island is and will henceforth be referred to by its nickname, 'Big Island'.

The current population of the state of Hawaii is approximately 1.4 million. The majority of the population lives on the island of Oahu. The Big Island has a population of 200,983 (U.S. Census Bureau, 2018). It is larger than all the other major islands put together, with 63% of the land mass but only has 14% of the state's population. Much of the island is considered rural and those rural areas are generally lacking public utilities including water. Most of the homes without municipal water are on the south side of the island (Figure 4.2).

Big Island is home to two large mountains (Figure 4.3), nearly 14,000 feet above sea level, and as a result, has very diverse rainfall and weather patterns. While some areas have very low average annual rainfall, the majority of people who harvest rain see an average of 32 to 133 inches of rain a year (Frazier *et al.*, 2016).

Figure 4.3 Hilo Bay. Mauna Loa on the left and Mauna Kea half hidden by clouds on the right. Both mountains are close to 14,000 feet above sea level and influence rainfall. (*Source*: Authors).

Figure 4.4 Old redwood tank located in the town of Volcano. (*Source*: Author).

4.1.2 Catchment history

Regular European contact with the islands started in the late 1700's and it was likely that water catchment became popular in the early 1800's. The older water tanks were usually wooden. A few old redwood tanks are still in use (Figure 4.4).

Today, individual homeowners with water catchment systems serving less than 25 people are the responsibility of the owner and there is no government oversight.

Wells exist but are costly and often impractical for a homeowner to dig because much of southern half of the island is made of layers of volcanic rock rather than soil. There is some surface water. Streams that do exist are usually intermittent or quite varied in their flow because the lava rock that makes up the island is so porous. For domestic water, catchment was and remains more practical.

In the 1960's subdivisions were set up as private subdivisions, meaning property owners were responsible for any utilities and/or roads. The subdivisions were often large. The largest is 36.8 square miles consisting of 157 miles of roads and about 11,500 lots averaging 1 acre each (HOVE, 2019). This particular subdivision, Hawaiian Ocean View Estates (HOVE), when full, will accommodate an average of 34,500 residents, without any municipal utilities. People were attracted by the idea they could get affordable land with few regulations other than basic county building codes. The owner had to figure out how to get water and could put in any kind of water system. Regulations for catchment still only require that a catchment system not be in the setback of neighboring properties and not be made of lead components. Only if there are specifics in a person's building plans for a catchment system, does the system have to conform with those specific guidelines. It is entirely up to the owner whether they store water in a horse trough or swimming pool, catch rain from a roof or the cow pasture.

As these water systems were private, public agencies, in particular the State Department of Health, took no responsibility in servicing that population. The result was having no government agency to turn to for accurate information and guidance. As late as the 1980's the Big Island's Department of Health representatives still refused to help individual homeowners with basic catchment water quality questions. That left residents on the Big Island with no one other than commercial vendors, friends or neighbors to advise them about their catchment systems. Today most subdivision lots have access to electricity, which makes rural living much easier. Surprisingly the lack of municipal water does not have a great effect on home prices. Catchment is considered a reasonable and viable alternative, and is sometimes preferred to municipal water. Another great change was within the Department of Health. While not happy about the subdivision being built without water in the first place, they are more responsive to people needing help with their catchment systems.

In spite of the challenges of the rural subdivisions, they are popular for people like first-time homeowners and retirees who can't afford town prices, for people who want space to operate a small business or farm, for those who like privacy and rural living, and more recently for those of greater economic means who want a place where they can build a big home on a large property. Rural single-family homes on catchment are also popular for people receiving county housing subsidies because the subsidy might otherwise only get them a small apartment in town.

4.2 UNIQUENESS OF THE BIG ISLAND CATCHMENT

One of the challenges of addressing water catchment issues on the Big Island is the wide diversity of systems. Since the properties are largely classified as agricultural/residential, rainwater systems can serve greenhouses, fruit and produce farms, livestock, and pets as well as homeowner needs. The communities are also home to churches, schools, community centers and private small businesses that are run from the home. This includes vacation homes (houses rented out for short term stays without an owner in residence), Bed and Breakfast establishments (where owners are in residence), regular rental homes, etc. Where whole towns are dependent on catchment – the town of Volcano for example – you can also add inns, restaurants, public restrooms, laundromats, fire departments, post offices, and so forth. This huge variety of water needs makes rainwater catchment solutions quite broad and diverse. There is also great variation in socio-economic status, age and education, which often defines the available budget.

At one time there was a stigma attached to living in an area without municipal services. As rural subdivisions became more attractive to all socio-economic levels and self-sustainability became popular, perceptions of living on catchment has become increasingly positive.

4.3 REGULATIONS

United States Environmental Protection Agency (EPA) has stringent rules for any water system that supplies water to more than 25 people. These regulations can be expensive and difficult for small businesses, private schools or churches to comply with. Rather than bear the burden, most find it easier and cheaper to pay for potable water to be trucked to their place of business and kept in storage tanks. Rainwater is still caught and used for the non-potable uses and is unregulated by the EPA. One beekeeper, who extracts and bottles honey on his property, uses treated catchment water for his home use and keeps a small tank for trucked-in municipal water for his honey processing. A community center trucks in water for kitchen use and drinking water, but uses catchment for flushing and cleaning. One laundromat collects rain in a small open reservoir for the washing machines after removing sediment. Public awareness of raw water risks has improved significantly in the past twenty years, but there remains a large public educational task.

Some larger facilities undertake the EPA regulations and treat their rainwater to potable-use standards. One of these is the Hawaii Volcanoes National Park which hosted close to 4 million visitors in 2017 (HVNP, 2019). The park not only collects, treats and uses rainwater for all their water needs, they also promote the sustainability of their systems by encouraging visitors to fill their water containers with treated rainwater from their taps (Figure 4.5).

Another large and unique facility following EPA regulations is the Kilauea Military Recreational Camp (Figure 4.6), a separate recreational camp primarily for military families that is within the National Park. It offers 90 one-, two- and three-bedroom cottages and apartments, dormitories, a restaurant, laundry, snack bars, theater, church, meeting rooms and catering (KMC, 2019). Rain is collected

Figure 4.5 National Park sign promoting the use of rainwater. (*Source*: Authors).

Figure 4.6 The entrance to Kilauea Military Camp.

Figure 4.7 A few of the larger potable water storage tanks for Kilauea Military Recreational Camp.

from a large roof surface designed specifically as a catchment surface and from the roofs of almost all the facility buildings. In total, they have about 6 acres of catchment surface area. For every inch of rain, they can collect about 166,000 gallons of water. The rainfall average is 65 inches per year. There are almost 3.5 million gallons of water storage available (Figure 4.7), and about 20,000 gallons of water are used per day. Rather than treating all the tanks to potable standards, only about 560,000 gallons of storage is for treated potable water. Other tanks have dedicated, non-potable uses such as firefighting, cleaning, laundry, etc., and therefore do not need to be treated to EPA standards (KMC, 2007).

4.4 WATER QUALITY

In 1981, a young woman bought her first home, a rural single-family home in southern Hawaii. The home's water tank was a portable, above-ground swimming pool with a soft permeable cover drooping into the water (Figure 4.8). The dragon flies were enjoying their artificial pond in the sagging center and the algae enjoyed a lovely bloom both outside and inside the tank.

Figure 4.8 Permeable ground cover cloth commonly used as tank cover. (*Source*: Authors).

With some concern, the woman called the Department of Health office on the Big Island and asked if there was something she should be doing to insure her water quality. One 20 micron sediment filter at the point of entry was the extent of her water treatment. The Department of Health agent told her that catchment systems were private and therefore not their responsibility. They made no suggestions or recommendations. Unaware of water quality issues, she spoke to friends and neighbors. The advice they gave her was: 'If the algae or plants growing in the tank are green, then the water is fine. You only have to worry if they start to turn brown'. Everyone else she spoke to, including store clerks selling filters, told her that 20–30 micron filters were all you needed, and that indeed seemed to be what everyone else that she spoke to was using. There were a few people buying drinking water but the general opinion was that it was just a taste preference and not really necessary.

When addressing rainwater quality issues, particularly when the water user is having problems, one has to evaluate a lot of variables. Even within a single subdivision there are micro climates. There can be huge variations in vegetation, (which can add a habitat or source of contaminants), socio-economic factors, education, legal requirements and treatment options. Of primary concerns are bacterial contaminates, for example *Salmonella* (an ingestion hazard carried by reptiles and birds) or *Leptospira* (an ingestion and contact hazard from mammal urine); *Protozoa*, such as *Giardia lamblia* (an ingestion hazard from feces of infected animals); parasitic worms such as *Angiostrongylus cantonensis* that causes rat lungworm disease (an ingestion hazard and possibly contact hazard); and chemical toxins. Viruses have not been observed to be a problem here yet but could be an issue in the future as exotic invasive species continue to arrive on the island.

Socio-economic status can effect what a homeowner can afford to do in terms of treatment. For example unlimited funds will buy a good whole-house treatment system with automatic reminders to change filters or monitor water levels. However, many of lesser means must find more cost-effective solutions. The central issue with treatment is people suddenly finding themselves as sole overseers of a critical system that they really don't understand, with risk factors they have never had to take into consideration. Even people who grew up on catchment can be misinformed and/or complacent about maintenance and treatment.

Rural Big Island often appeals to people with an independent spirit who like living semi-independent of government services. They often get quite inventive with treatment designs and water systems. Some are quite clever, for example using nylon hose for pre-tank large-sediment filters. Others can be misguided, such as the man who used rainwater to fill a pool used for swimming. He then used the same swimming water for his domestic supply. He treated it heavily with chemical disinfectants, not aware of the associated health risks of drinking swim water, nor making the connection between those harsh chemicals and the failure of several water heaters.

A big issue is people who have no treatment and may have done fine without it for decades. It's often hard to convince them that they need water treatment, particularly since risks have increased with the population growth and influx of exotic vectors and diseases. Even without new threats, these residents often don't understand that although they may have no ill effects from their water because they've developed immunities to its unique microflora and common contaminates, others not so immunized, such as a visiting grandmother, may suffer adverse effects. One sad incident occurred when short term/emergency foster children were placed into care in a home that had a catchment treatment system, but sadly the ultraviolet light in their system was no longer working. The foster family was not ill, having built up immunities to their water, but unfortunately the foster children, unused to the tank's bacterial biota, became ill. Inspectors who may have noted that there was an adequate treatment system at the house likely had no means or awareness to make water testing part of the placement process.

4.5 RAINWATER CATCHMENT PROGRAM BEGINNING

By 1999, the population had grown substantially in the rural neighborhoods. Complaints from residents getting ill from their water were getting more public attention. Only two studies on illness related to the use of water catchment had been done on the island. One study in 1993 linked the probability of getting *Leptospirosis* (a bacterial disease passed through contact with mammal urine) with catchment use (Sasaki *et al.*, 1993). The other study was less formal and happened in the 1990's when national attention was drawn to people on catchment developing lead poisoning. Lead paint was sometimes used in older plantation homes to coat roofs and some used lead-headed nails to secure metal roofs. Some residents even lined the inside of their wooden tanks with lead paint to keep algae growth down. Without any oversight or knowledge, homeowners had little information to learn of the danger in which they had put their families.

Catchment tank vendors were the ones most people turned to for information on how to build and maintain a water system that didn't risk the health of the water user. Local businesses were primarily geared to selling and setting up tanks, pumps and water conveyance systems, not in toxicology or microbiology. Nevertheless, many vendors were very conscientious about providing good systems for their clients. Even with conscientious island vendors, there was little information available especially information relevant to island catchment needs. Mainland suppliers were not always aware of the unique issues. For example, treatment systems which were designed to improve pre-treated municipal water were sometimes sold for the raw (untreated) water of catchment users and obviously not up to the task.

In 1999, research on how to assist homeowners understand the threat of *Leptospirosis* was begun by the author, who worked in the University of

Hawaii's Cooperative Education program, leading to the start of a statewide rainwater catchment program (RCP).

A community survey was initiated to see if catchment users understood what *Leptospirosis* was, where it came from and if they thought they were at risk. It also asked general questions about the ownier's systems and maintenance. Responses pointed to a glaring lack of basic knowledge about how *Leptospirosis* was contracted, water quality in general, water-system maintenance and water treatment. The survey responses also demonstrated that there was a glaring lack of information available to the homeowners so they could learn more. Stunningly, not one person surveyed identified *Leptospirosis* as a contact hazard and not just an ingestion hazard.

To compound the lack of information available, there was only one laboratory at this time where people could get their water tested. That lab was 2–3 hour drive away for the majority of residents, a 4–6 hour round trip. Shortly thereafter that lab closed leaving no island options for individual water testing. State Department of Health labs do not perform tests for private water systems.

4.6 THE RAINWATER CATCHMENT PROGRAM

An important goal of the rainwater catchment program (RCP) was to enable residents to help themselves. To do this it needed to provide residents with good, comprehensive information. The RCP started with the production of informative guidelines for residents. The guidelines (Macomber, 2010) gave answers to essential catchment system questions, and the maintenance of those systems from raindrop to faucet. Funding from an NGO and the State Department of Health helped publish the guidelines so that they could be distributed without cost to state residents. The result was the nation's first statewide rainwater catchment system guidance. Over 40,000 have been distributed in hard copy form and continue to be available digitally.

As residents had no way of knowing if their water systems were contaminated or not, or if their water treatments worked, the second priority of the new RCP was enabling residents to monitor their own water quality. Simple, inexpensive test kits for fecal indicators (Figure 4.9, (Source: Authors)) were designed by the RCP using the Manja method (Manja *et al.*, 1982). These kits were then distributed or sold at cost through the University Extension office and later by local vendors who volunteered to assist in their distribution.

Initially testing was done at the University's Cooperative Extension office (Figure 4.10; Source: Authors), but as soon as a home test kit could be provided rural residents could easily pick up tests kits when they were in town and have them available when needed. This enabled the catchment owner to finally be able to manage their own systems, and know if what they were doing was successful. It also freed up the single RCP personnel to continue with other catchment

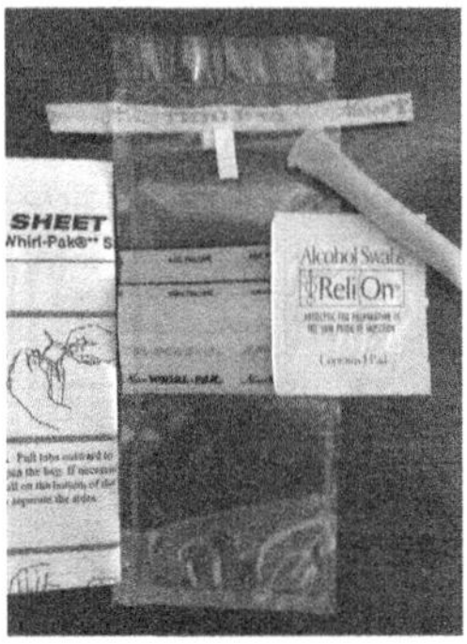

Figure 4.9 Home test kit.

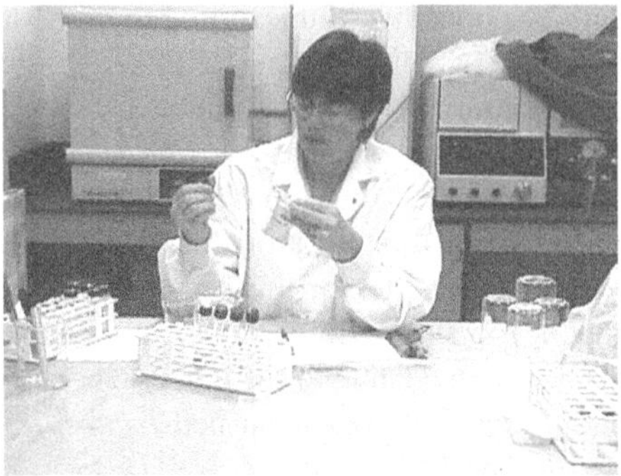

Figure 4.10 RCP water tests.

outreach work since the popularity of the testing was immediately quite overwhelming and time consuming.

The Manja method was chosen because it used hydrogen-sulfide producing gut bacteria as indicators rather than the typical tests for *E. coli*. There are indigenous, non-gut *E. coli* found in the Islands which could lead to false positive results if using typical *E. coli* indicators (Hardina and Fujioka, 1991). Also, the Manja formula was less sensitive to temperature variations and could be used without incubation in lower elevations where temperatures were warm. This meant that a catchment user could do this test themselves at home (Rijal and Fujioka, 1995). One other note in favor of the Manja method testing is that the test results, if positive, encourages action by the homeowner due to its foul odor and color change from clear yellow to opaque black. This putrid ugly result has been known to convince even the adamant denier into cleaning his system. Once the program became established, the community need for it was easily confirmed and funding to expand services was easier to obtain.

4.7 RAINWATER CATCHMENT PROGRAM EXPANSION

4.7.1 Education

Education was always a key goal of the catchment program. Besides the guidelines, frequent community workshops explaining the 'problems and resolutions' for safe catchment use were offered throughout the island. Eventually, specialized workshops were offered (Figure 4.11). Topics included how to repair and troubleshoot pumps, how an ultraviolet light system works and other directly relevant subjects. Many simple handouts (Figure 4.12) were also created to simplify some topics, meet some specific needs and reach out in other languages for those with limited English. Seminars were provided to realtors, home inspectors, business owners, government agencies, businesses and other

Figure 4.11 Pipe fitter Rick Bishop explaining pump repair.

Figure 4.12 RCP Brochure examples. (*Source*: Authors).

specialized groups that needed to know more about catchment systems. Manned informational booths were set up in county fairs, community fairs, health fairs, and sustainability fairs. Local radio stations invited RCP to speak. Training was delivered for outreach workers in health fields, particularly those dealing in public health.

4.7.2 Available agent

One challenge worth mentioning is the fact that systems were quite individualized. Homes and businesses ranged from shacks to castles and water tanks and systems were just as diverse. There were few standards and the population consisted of many very self-reliant and creative problem solvers. There was rarely a once-size-fits-all answer to homeowner's questions. Even with comprehensive guidelines and seminars, having a person available to answer individual questions rapidly became an essential component of this valued pubic service, requiring more office hours and a message system.

4.8 COLLABORATIONS

4.8.1 Vendor support

Including and involving local vendors in the RCP was beneficial to all concerned. Vendors were often the first contact for new catchment owners or those with problems. Vendors benefited from free University literature to hand out, and inexpensive test kits to sell, increasing their public image and providing a service for their clientele. For the RCP, to have distribution of educational materials to a

Figure 4.13 Tradeshows were great outreach tools. (*Source*: Authors).

targeted audience helped spread awareness of water quality maintenance and risks. The RCP provided vendors and catchment users a place to call for questions and for more-sophisticated water testing.

Many vendors agreed to participate at community trade shows (Figure 4.13) promulgated by the RCP to service the catchment clientele. This allowed rural residents to make one trip for comprehensive information on various options available to them. It allowed vendors to show off their wares and services to potential customers. Finally, it allowed the RCP to display educational models and distribute materials to a targeted audience.

Working hand-in-hand with vendors was a great advantage to the RCP, the vendors and the catchment community. The RCP made it clear that it would not favor or promote any particular vendor over another, but tried to encourage the homeowner to discuss options and technologies with all of them. An interesting development of this partnership was that as emphasis on water quality was raised by the RCP, the sales of treatment systems began to outsell the water storage tanks, the previous focus of the catchment stores. This was a boom for the vendors which encouraged further collaboration, benefitting all involved.

4.8.2 Expanding research and collaborations

Scientific studies were important in determining and responding to program direction and consumer needs. Collaboration with experts in other fields helped bring in research grants. Collaborative programs were initiated in other US protectorates and territories where similar water quality/quantity problems exist.

For example, Train-the-Trainer workshops, conferences and specific programs such as introducing low-tech water quality improvement options, were run in various states in four Micronesian countries.

4.9 ENVIRONMENTAL CHANGE EFFECTS ON RAINWATER CATCHMENT

Understanding environmental changes and how they affect catchment systems is another area that needs more research and collaborations.

4.9.1 Weather/rainfall

Weather changes have brought challenges to residents. Droughts might be hard to imagine for a tropical island, however due to wind patterns, the topography of the Big Island includes deserts where rainfall is limited or intermittent. Residents in these areas often have larger catchment surfaces and tanks to get through the dry periods, however prolonged dry spells can still mean low water supplies. Where rainwater is usually plentiful, unexpected droughts can be a major problem because water storage and catchment surfaces are much too small. In addition, residents used to abundant supplies are not used to checking water levels in tanks and may be surprised by a 'drought'. It is common even after two weeks without rain on the rainy side of the island to hear a steady stream of water tanker trucks passing by to fill small tanks that have already run out of water. After three weeks without rain, the waiting time to get a truck available to deliver water gets long. A month long 'drought' is an emergency. Another instance of surprise is when there is fairly constant, but very light rain. This situation can lull residents into thinking there is plenty of water if tanks are not checked. Another unfortunate side effect of an empty tank is that 'on-demand' pumps get damaged when the water levels drop below the intake pipe. Climate change appears to have increased the variability of rainfall, making rain consistency less reliable.

All these issues can be addressed with increased conservation or greater storage and/or catchment surface, if one can afford it. But it is also important that residents adopt behavior patterns that make them aware of the weather, rainfall quantities, tank water levels and the need to conserve when necessary.

The county has helped considerably to make potable water available by getting a grant to install public water spigots (Figure 4.14) where rural residents could fill containers up to 50 gallons/day. There are currently 18 public spigot locations available on Hawai'i Island that supply residents with free water during droughts and year-round potable water. The spigots have been well used and much appreciated, the only downside being that they can only be installed where municipal water is available, which sometimes that means residents still may need to travel many miles to get water.

Figure 4.14 Resident getting water at County spigot. (*Source*: Authors).

Figure 4.15 Semi slug. (Photo coutesy of K. Howe).

4.9.2 Environmental change: Invasive species

Some environmental changes that the rainwater catchment program works to address are caused by invasive species. A recent example of a threat to water catchment users is the invasion and subsequent overpopulation of the semi-slug, *Parmarion martensi* (Jarvi *et al.*, 2018) (Figure 4.15). The semi-slug is an invasive species to Hawai'i that is an extremely efficient carrier of a parasitic worm *Angiostrongylus Cantonensis* which causes a disease known as Rat Lungworm. One of the characteristics of the semi-slug is that it likes to climb (Hollingsworth *et al.*, 2007), particularly towards moist places. Observations of semi-slugs climbing up and into water tanks are common. With the soft covers on the majority of the tanks on the island, there are often gaps where pipes go under the cover that make a perfect thoroughfare for the slug to get access to the tank. The semi-slugs often line the sides of the tank then drown as water levels rise with the rain, as they fall or get washed in. The microscopic parasitic worm escapes the drowned semi-slug and has survived 21 days in fresh water at the University of Hawai'i at Hilo School of Pharmacy's research lab (Howe *et al.*, 2019). Infections of the *A. cantonensis* can cause moderate to severe brain damage to those who ingest the parasites, the damage relative to the quantity of parasitic worms entering the body (Jarvi *et al.*, 2018). Since the infectious worms have been observed migrating to the bottom of the water column (Howe *et al.*, 2019) where the water intake valves are normally located, it is assumed that the chance of getting a high dose of parasites from an infested tank is possible. New studies from the University of Hawai'i Hilo School of Pharmacy show that treatment systems previously thought sufficient may not be adequate for blocking this new waterborne threat (Howe *et al.*, 2019).

Unfortunately the spread of the semi-slug isn't limited to the tropics. Climate change is spreading rat lungworm to other US locations where they have not been a problem before. For example, reports from Texas suggest

Figure 4.16 Puu Oo vent. (Photo courtesy of Hawaii Aerial Visions LLC).

that the semi-slug may be responsible for three cases of rat lungworm where people came into contact with parasite infested flood waters (Foster *et al.*, 2016; Hammoud *et al.*, 2017).

4.9.3 Environmental change: Volcanic

One unique environmental concern for catchment users in Hawai'i has been volcanic emissions (Figure 4.16). Hawai'i has had one fairly continuous volcanic eruption from 1983 through 2018 and numerous others. Even without an active outbreak of lava, there are still fumes, primarily sulfur, being emitted through fissures. Sulfur emissions create particularly acidic rain, which is one of the problems for catchment users. The more acidic the water, the more leaching there could be of all the collection, storage and distribution components. Another problem with volcanic eruptions is explosive events that put large amounts of particulate matter into the air, causing wide dispersal of heavy loads of contaminants.

Each eruption produces a different composition of gases and particulate matter that varies greatly in quantity, causing equally varied problems and solutions. While air, dust, rain, fumes, ash and particulates can be measured during an eruption, and assumptions can be made on its effects and toxicity, there had been no studies on the cumulative effect on catchment water quality over a period of time. This concerned the RCP office and triggered a study.

4.9.4 Environmental study

In 2009, an eruption produced an exceptionally large amount of emissions. A collaborative study among the Center for the Study of Active Volcanos, the County of Hawai'i, and the RCP was undertaken because of concerns over water quality in catchment tanks. There was concern that these emissions, which included sulfur dioxide and sulfur trioxide being discharged at 700–1100 metric tons per day, would affect the acidity of collected rainwater. Hydrofluoric acid and hydrochloric acids were also present in the plume and both can form aerosols that will deposit on roofs and wash into the water supply. This caused concerns that levels of fluoride would exceed EPA maximum contaminate levels (Thomas & Macomber, 2010).

The team collected samples from 439 catchment tanks in 3 communities giving the residents immediate feedback on their pH levels and neutralization methods. In two communities downwind of these acidic fumes, the median pH of untreated catchment tank waters was 4.3–4.4. Lower extremes were 2.9–3.3. (Thomas & Macomber, 2010). Because of the RCP, the public was continuously informed by e-mail (or letter) of their individual water test results along with interpretations and suggestions for remediation action if needed. They were also given their overall community results, and comparative results (and conclusions) of all communities involved were shared with everyone. This type of feedback gave the participant a reward for their efforts and some ownership due to their participation. It paved the way for obtaining future contributions. Information was also supplied to community centers (and government officials) where it could be shared with non-participant residents to the benefit of the overall health of the community.

While collecting samples, data was also collected to obtain an overview of water availability, tank size, treatment techniques and systems, effects of wind on particulate distribution, and, surprisingly, attitudes of people with regards to water. For example, those living where rainwater was most abundant were much more lackadaisical in their water quality oversight. Those living in drier areas were much more vigilant and active in caring for their water supply. This information is very useful in directing the future of rainwater catchment programs.

4.10 SUMMARY

Harvesting rainwater is an ancient practice that can be adapted successfully to modern lifestyles. From the early days of plantations and redwood tanks, Big Island rainwater catchment usage has become an essential part of the lifestyle. Homes, subdivisions and towns steadily increase in size and often with that growth, comes a greater dependence on rain. Larger populations living closer together, introduced vectors, and invasive species mean challenging new threats to captured water quality. In spite of water quality challenges harvesting rain has proven to be a viable and sustainable water source.

Technological advances have offered more options for water treatment, but before treatment can be effective, one must understand the uniqueness of the local rainwater resource, the catchment system characteristics and the specific problems, and therefore the specific solutions needed to supply healthy water in sufficient quantities. The overall goal of the Hawai‘i RCP has been to increase education, services, knowledge and awareness of water quality and quantity to enable residents to be competent stewards of their own water systems. Comparative surveys show a significant increase in residents using better systems and sufficient water treatment compared with before the RCP. Vendors have verified the changes in the community, noting revenue from treatment systems has grown and now exceeds that from water tank sales. These and other benchmarks are some of the verifications of a successful program.

In spite of continued challenges, such as climate change and volcanic emissions, rainwater catchment on the Big Island of Hawai'i will remain a major source of potable and non-potable water for large segments of the population for the foreseeable future. Although private water systems are not governed by the US Environmental Protection Agency, consumption of rainwater by large portions of the population is nevertheless a public health issue that should be fully addressed by the state and particularly county government. Without a dedicated champion or team and sustained funding, programs such as the Rainwater Catchment Program described here will soon fail to deliver the needed guidance and public health will suffer.

REFERENCES

Foster C. E., Nicholson E. G., Chun A. C., Gharfeh M., Anvari S., Seeborg F. O., Lopez M. A., Campbell J. R., Marquez L., Starke J. R. and Palazzi D. L. (2016). *Angiostrongylus cantoneensis* infection: a cause of fever of unknown origin in pediatric patients. *Clinical Infectious Diseases*, **63**, 1475–1478, https://doi.org/10.1093/cid/ciw606 (Accessed: May 2019).

Frazier A. G., Giambelluca T. W., Diaz H. F. and Needham H. L. (2016). Comparison of geostatistical approaches to spatially interpolate month-year rainfall for the Hawaiian Islands. *International Journal of Climatology*, **36**, 1459–1470, doi: 10.1002/joc.4437 http://rainfall.geography.hawaii.edu/howtocite.html (Accessed April 2019).

Hammoud R. A., Nayes S. L., Murphy J. R., Heresi G. P., Butler I. J. and Perez N. (2017). *Angiostrongylus cantonensis* Meningitis and myelitis, Texas USA. *Emerging Infectious Diseases*, **23**(6), 1037–1038. https://doi.org/10.3201/eid2306.161683.

Hardina C. M. and Fujioka R. S. (1991). Soil: The Environmental Source of *Escherichia coli* and *Enterococci* in Hawaii's Streams. *Environmental Toxicology and Water Quality* **6**, 185–195.

HFD. (2008). County of Hawaii Fire Department via The Hawaii Tribune Herald. Hilo, Hawaii. Redrawn digitally by author.

Hollingsworth R. G., Kaneta R. K., Sullivan J. J., Bishop H. S., Qvarnstrom Y., DaSilva A. J. and Robinson D. G. (2007). Distribution of *parmarion cf. martensi*, A New Semi-Slug Pest on Hawai'i Island, and its Potential as a Vector for Human *Angiostrongyliasis*. *Pacific Science*, **61**, 157–168.

HOVE. (2019). Hawaiian Ocean View Estates, https://hoveroad.com/community-links (Accessed: April 2019).

Howe K., Kaluna L., Lozano A., Torres Fischer B., Tagami Y., McHugh R. and Jarvi S. (2019). Water transmission potential of *Angiostrongylus cantonenesis*: Larval viability and effectiveness of rainwater catchment sediment filters. *PLoS One* **14**(4), e0209813, https://doi.org/10.1371/journal.pone.0209813

HVNP. (2019). Hawaii Volcanos National Park, https://irma.nps.gov/Stats/SSRSReports/Park%20Specific%20Reports/Monthly%20Public%20Use?Park=HAVO (Accessed: April 2019).

Jarvi S. I., Howe K. and Macomber P. (2018). Angiostrongyliasis or rat lungworm disease: a perspective from Hawai'i. Current Tropical Medicine Reports Tropical Parasitology (A Sanchez, Section Editor). Springer International Publishing AG, part of Springer Nature. Published online: 05 February 2018 https://doi.org/10.1007/s40475-018-0132-z

KMC. (2007). Kilauea Military Camp, https://kilaueamilitarycamp.com/ (Accessed: April 2019). Personal communication and site tour by KMC water maintenance personnel.

KMC. (2019). Kilauea Military Camp, Recreational facilities, https://kilaueamilitarycamp.com/ (Accessed: April 2019).

Macomber P. (2010). Guidelines on Rainwater Catchment Systems for Hawaii. Mänoa: Collage of Tropical Agriculture and human Resources, University of Hawai.

Manja K. S., Maurya M. S. and Rao K. M. (1982). A simple field test for detection of faecal pollution in drinking water. *Bulletin of World Health Organization*, **60**(5), 797–801.

Rijal G. and Fujioka R. S. (1995). A Homeowners Test for Bacteria in Cistern Waters, Water Resources Research Center. University of Hawaii at Manoa, Honolulu, Hawaii.

Sasaki D. M., Pang L., Minette H. P., Wakida C. K., Fujimoto W. J., Manea S. J., Kunioka R. and Middleton C. R. (1993). Active surveillance and risk factors for *Leptospirosis* in Hawaii. *American Journal of Tropical Medicine Hygiene*, **48**, 35–43.

Thomas D. M. and Macomber P. S. (2010). Impact of Persistent Degassing of Kilauea Volcano on Domestic Water Supplies. American Geophysical Union. Presented 13-17 December 2010 American Geophysical Union, Fall meeting, San Francisco, CA USA. Also presented 14-17 September 2009, American Rainwater Catchment Systems Association Conference, Decatur GA USA.

U.S. Census Bureau. (Data released on December 19, 2018) http://dbedt.hawaii.gov/economic/datawarehouse/ (Accessed: April 2019).

Chapter 5

Mexican rainwater harvesting movement in recent years

J. A. Gleason Espíndola[1], Y. Corona Sánchez[2,3] and C. Casiano Flores[3]

[1]University of Guadalajara, Urbanism, Guadalajara, Mexico (arturo.gleason@cuaad.udg.mx)

[2]Tehnological Water Research Institute Lic. Arturo Gleason Santana, Guadalajara, Mexico (Yahaira.corona@iitaac.org.mx)

[3]Public Governance Institute, KU Leuven, (cesar.casiano@kuleuven.be)

Keywords: certification, floods, Mexico, rainwater, harvesting, history, scarcity, water supply

5.1 RAINWATER HISTORY IN MEXICO

In Latin America and especially in Mexico, rainwater harvesting (RWH) has been a technique developed and practiced since before the Spanish conquest. The Mayan and Aztec cultures captured and distributed rainwater using channels for drinking and irrigating their crops during the dry season.

Mexico has invaluable experience with rainwater harvesting. The Mayan "Chultuns" (Figure 5.1) used since the late pre-classic period was a common practice to save water for crops and supply water for people. The Chultuns were underground excavations waterproofed with gesso (González De la Mata, 2003). In Oxkutzcab at the skirt of Puuc's mountain, the water was captured at an area of 100 to 200 m^2 and stored in Chultuns for later distribution (McAnany, 1993).

Unfortunately, all the advances for the implementation of rainwater harvesting were used less frequently after the Spanish conquest. Mostly the rainwater

Editor(s): José Arturo Gleason Espíndola, César Augusto Casiano Flores, Raul Pacheco-Vega and Margarita Rosa Pacheco Montes
doi: 10.2166/9781789060584_0073

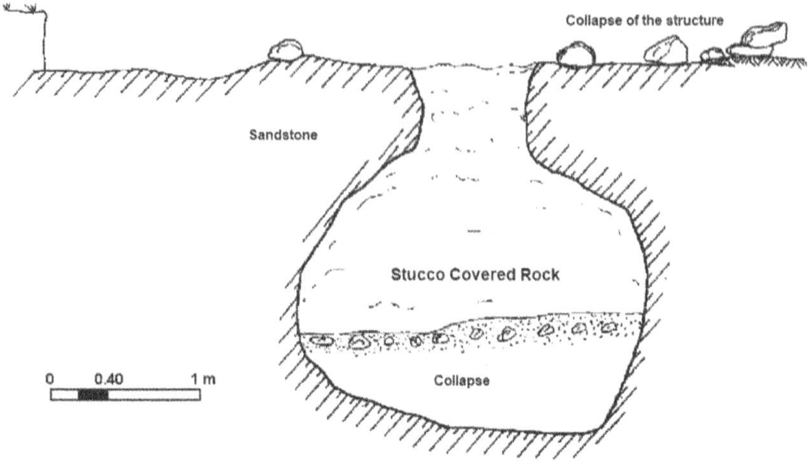

Figure 5.1 Chultuns in mayan culture. (*Source*: González De la Mata, 2003).

harvesting systems were replaced by the construction of aqueducts, wells, and Spanish agricultural practices.

Another factor that helped reduce the use of rainwater was population growth. Between the nineteenth and twentieth centuries, most of the world had dramatic population growth. As a consequence, the demand for water skyrocketed. This large amount of water was and even now is supplied by aqueducts and centralized water supply systems. However, in recent years, rainwater harvesting is gaining interest again worldwide. Rainwater is used not only to store water for dry periods but also as a measure to cope with extreme rainfall (Hofman & Paalman, 2014).

5.2 RAINWATER IN 21ST CENTURY MEXICO

Every day the water supply sources get more affected by overexploitation, poor management, and climate change. As a consequence, millions of people suffer from a lack of supply of good quality water (UN, 2018). As a result, water takes greater relevance in societies, whether there is not enough for basic needs or because water pollution causes damage to health or floods.

In Mexico, most of the water consumed comes from a threatened lake, river, or an overexploited aquifer (CONAGUA, 2015), both requiring an enormous amount of energy (WWAP, 2015). Therefore, rainwater harvesting is regaining power as an alternative water supply. Among the most representative efforts of the implementation of rainwater harvesting is in Mexico and one of the pioneers of this technique is Dr. Manuel Anaya Garduño, a professor from the Graduate School (COLPOS) in Texcoco, State of Mexico. He designed a rainwater

Figure 5.2 Tlaloque Urban Island. (*Source*: Isla Urbana, 2019).

harvesting system to help solve the problem of drinking water for the benefit of more than 150,000 rural people in communities that lack infrastructure. The system consists of catching the rainfall from roofs, storage in tanks, and conveyance to a purification plant, where the water is purified, and used in the community.

Dr. Manuel Anaya is also the leader of the academic rainwater harvesting systems network, sponsored by the Science and Technology National Council (CONACYT). This network has the objective of exchanging ideas, experiences, and innovations of rainwater harvesting systems to contribute to solving water scarcity problems, especially in marginalized areas.

To the efforts of Dr. Anaya are added the creation of Urban Island (Figure 5.2), a Non-Governmental Organization focusing on the south of Mexico City. The municipalities of Tlalpan, Iztapalapa, and Xochimilco live in a water paradox of water scarcity and floods.

Frequently houses of this area do not receive water supply for weeks, receive poor water quality, or do not have water services at all. The lack of a good water supply forces families to purchase water carried by truck, this can become very expensive and unaffordable for people in the area, and still be very irregular.

Figure 5.3 RWH in Mexico City. (*Source*: Sedema, 2019).

Figure 5.4 Tarahumara's RWH. (*Source*: Captar A.C 2019).

Consequently, Urban Island has been working in those municipalities installing rainwater harvesting systems to supply water to those families without water services. Tlalpan and Xochimilco are two of the most impacted communities, with almost 4,000 rainwater harvesting systems installed by 2018, and more than 23,000 benefited people (Isla Urbana, 2019).

Another sample of the rainwater harvesting evolution in recent years is the case of San Felipe, a community located in the State of Guanajuato, Mexico, at an altitude of 2,140 meters above sea level, with high temperature and desert-like climate. The "Water and Life" project began with the implementation of tanks in 1996. The first technological development consists of a rainwater system with a storage capacity of 500,000 liters and a catch covered by stone.

In some places, the struggle caused by water scarcity and poor water service creates a severe crisis in the population. For example, in 2005, a drought attacked Ocotlan's Valley in Oaxaca, and for over ten years, the people watched their crops die because of lack of water. The water crisis in this indigenous community made them look for alternatives to survive, that's how nowadays hundreds of farmers harvest and infiltrate the rain every year. Using the rainwater during the dry season, the farmers grow fruits, vegetables, and roses (EFE, 2019).

Social initiatives have been emerging to supply water where homes suffer from severe scarcity or/and inadequate water quality. NGO's and sometimes the local government join forces to reach the maximum number of people possible. That is the case of the rainwater harvesting (RWH) program recently launched in Mexico City. The Environmental Secretariat and Isla Urbana (Urban Island) together started a plan to install 10,000 RWH systems (Figure 5.3) in areas with severe water scarcity, especially in Xochimilco and Iztapalapa. This initiative aims to improve water access conditions, reduce floods, and save money and energy to the families.

Another excellent example of the expansion of good practice in the country is the work of Captar A.C, focusing on the wellbeing of the Tarahumara's indigenous community (Figure 5.4) located in the northern state of Chihuahua. Captar seeks solutions that influence fundamental problems for the people in the zone. The realization of technical and demonstrative workshops are an essential part of the process with the community. Since 2006, Captar has been installing rainwater harvesting systems, creating family farms, etc. They aim to help and work with people living in extreme poverty to create good quality and long term systems.

However, to discuss water problems, it is necessary to explore causes, to establish appropriate solutions based on multidisciplinary approaches. As a consequence, the University of Guadalajara (UDG), through the academic group Management and Technology for Architecture and Sustainable Urbanism and the Sustainable Urban Architectural Technology Laboratory (LATAU), were pioneers developing prototypes of RWH in households and universities campuses. Besides, the University generates data through a network of weather stations located all around Guadalajara's Metropolitan Area in collaboration with the Technological Water Research Lic. Arturo Gleason Santana (IITAAC). Since 2013, IITAAC has been promoting theoretical and applied research projects, programs that provide concrete answers to the challenges in water issues that overwhelm society and the environment.

IITAAC is a Non-Profit Organization seeking to promote water culture around the community through environmental education focusing on children and students. Using the Stormhunter truck (Figure 5.5), IITAAC can visit communities, schools, businesses, parks, etc., to share science and water education with the community, such as rainwater harvesting systems.

Figure 5.5 IITAAC's Stormhunter. (*Source*: IITAAC, 2017).

Figure 5.6 Sky Ha' Rainwater. (*Source*: Sky Ha, 2018).

5.3 RAINWATER AS AN INDUSTRIAL APPROACH

By 2030, according to the UN, total global water demand is expected to exceed supply by 40%, and approximately half of the world's population will suffer from water stress. Agriculture followed by industry is the most significant freshwater users, and the demand for water in manufacturing is expected to increase by 400% by 2050 (WWAP, 2015).

Sometimes rainwater can be seen just as a domestic or agricultural water supply. On the contrary, rain can be used as a part of the production chain or even as a raw material.

Sky Ha' Rainwater is the perfect example of rainwater as a raw material. Sky Ha' is a social and ecological enterprise that was founded to create natural products that promote the well being of humans and nature. All this is achieved by maintaining the highest quality standard and making use of sustainable technology and responsibility. Sky Ha' (Figure 5.6) is a company that catches, purifies, bottles, and distributes rainwater in hotels around the Mayan Riviera in southeast of Mexico.

In contrast, different companies like and Sistemas Pluviales in Mexico City, or Xocalli in Morelia are looking to develop integrated water management projects. Soluciones Hidropluviales (Figure 5.7) is a Mexican company focused on designing and developing projects involving rainwater and stormwater management. The company is a pioneer in generating and applying solutions for large urban catchment areas such as industries, shopping centers, housing developments, or municipal projects.

5.4 INSTITUTIONAL EFFORTS

5.4.1 Expansion of good rainwater practices

Population growth, water scarcity, lack of good water quality, and floods are some of the problems Mexican society faces every day. As mentioned in section 5.2 of this

Figure 5.7 Rainwater management in a public space. (*Source*: Soluciones hidropluviales, 2019).

chapter, in recent years, the rainwater harvesting movement has been skyrocketing around Mexico.

In the past 20 years, there have been different efforts to create a national collaboration to promote RWH regarding planning, development, management, and education. It all began with Dr. Manuel Anaya efforts at the 11th Rainwater Harvesting Systems International Conference in 2003, organized by IRCSA-Mexico.

The discussion, planning, and regulation are a crucial part of every practice and technology development. In 2017, the National Rainwater Catchment Systems Association (AMSCALL) was founded to promote sustainable rainwater practices and help restore the water cycle's health. AMSCALL seeks a national presence that generates opportunities for RWH, decreasing water shortages and floods.

In 2017 AMSCALL with the collaboration of IRCSA, the National Housing and Sustainable Communities Laboratory, University of Guadalajara (UDG), and IITAAC organized the First National Conference of Rainwater Harvesting Systems (CONAMSCALL by its initials in Spanish Congreso Nacional de Sistemas de Captación de Agua de Lluvia) in Guadalajara, Mexico. The 1st CONAMSCALL was a big international event and had a great attendance of people from all around Mexico and speakers from more than seven countries.

The institutional efforts to create a space where water professionals and people interested in rainwater can have discussions on planning, promotion, or regulation is essential for the development of the movement. That is why in 2019, AMSCALL presented the 2nd National Conference, this time in Morelia, Mexico. This time the main subject was Rainwater Harvesting for a Water Sensitive Mexico. The conference aimed to promote water management in cities and communities, oriented to the management and conservation of the water cycle, as well as to the optimal functioning of its urban hydrosanitary system. Rainwater harvesting becomes crucial in this process since precipitation is the entrance of water to the basin.

The 2nd CONAMSCALL focused on researchers, academics, architects, urban planners, engineers, as well as related professionals, students, NGOs, business persons, officials, government staff, the international community, and the general public interested in participating in rainwater harvesting.

5.4.2 Certification program to accredit professionals in rainwater harvesting

In past years, the implementation of Rainwater Harvesting Systems has been growing around all Mexico with the idea of increasing water for domestic, agriculture, and industrial supply. Furthermore, RWH has allowed a reduction in groundwater withdrawals and decreased stormwater volume.

For that reason, during the National conference of 2019, AMSCALL launched the National Certification Program for all businesses and people interested. All water professionals interested in rainwater should be evaluated every two years to continue with the certification.

The certification program comprises four levels:

(1) *Introduction to Rainwater Harvesting and Use: Promoter Certificate.*
This first program is aimed at those who are new to the rainwater world and requires basic knowledge about the water cycle and hydrosanitary systems.

(2) *Design of a Rainwater Harvesting System: Designer Certificate.*
This course is designed for building professionals or for those who wish to design a RWH system and aspire to develop a professional career in rainwater management.

(3) *Development of the Rainwater Harvesting System: Developer Certificate.*
Designed to provide knowledge about technical standards, pipes, design standards, calculations, installation of commercial and industrial systems, planning, and construction. Also, pumping systems, treatment systems, and practical application in the development of rainwater harvesting systems.

(4) *Supervising the Installation and Operation of Rainwater Systems: Supervisor Installation and Maintenance of Rainwater Harvesting Systems.*
This course aims to prepare professionals to supervise the installation and operation of rainwater harvesting systems by providing maintenance services. This course addresses professional skills to evaluate each component of systems to ensure its proper functioning.

The certification program content is based on the information from IRCSA, and the professional program created by The American Rainwater Catchment System Association. Additionally, every step was supervised and approved by the National Laboratory of Sustainable Households and Communities from CONACYT.

5.5 CONCLUSION

Historically, Mexican culture has had a strong relationship with water since the ancient prehispanic civilizations, developing rainwater harvesting techniques in the 10th century BC. Nowadays, the Mexican rainwater movement is getting stronger all around the country. However, it is still necessary to encourage higher participation in rainwater harvesting practices to build an extensive sector with a strong AMSCALL through the homologation, training, and affiliation of new members. Furthermore, to create consistent, reliable, and regulated systems, all the technological advances, and local experiences must be turned into technical rules that warranty an adequate system installation and function.

It is a significant challenge for the rainwater harvesting culture to be accepted in all sectors of the Mexican society and to settle on the basis of a transformation for sustainable water management. Finally, it is imperative to recognize all the people, NGO's, businesses, and universities who have been interested in promoting these significant subjects of water management and conservation.

REFERENCES

Captar A.C. (2019). *Centro de acopio para la Tarahumara A.C.* Obtenido de Captar.org: https://captar.org/impactos/

CONAGUA. (2015). Estadísticas del Agua en México 2015. CONAGUA, México.

EFE. (21 de marzo de 2019). *La sequía obliga agricultores a "sembrar agua" en el estado mexicano de Oaxaca*. Obtenido de EFE: https://www.efe.com/efe/america/mexico/la-sequia-obliga-agricultores-a-sembrar-agua-en-estado-mexicano-de-oaxaca/50000545-3931616

González De la Mata R. (2003). Los Chultunes de Chichen Itza. In: *Simposio de Investigaciones Arqueológicas en Guatemala*, Museo Nacional de Arqueología y Etnología, Guatemala, pp. 994–1008.

Hofman J. and Paalman M. (2014). Rainwater Harvesting, a Sustainable Solution for Urban Climate Adaptation? KWR, Nieuwegein, The Netherlands.

IITAAC. (2019, julio). www.iitaac.org.mx. Retrieved from Instituto de Investigaciones Tecnológicas del Agua Lic. Arturo Gleason Santana A.C: http://www.iitaac.org.mx/proyectos/cazatormentas/

Isla Urbana. (2019, julio). www.islaurbana.org. Retrieved from http://islaurbana.org/sistemas-de-ciudad/

McAnany P. (1993). Water Storage in the Puuc Region of the Northern Maya Lowlands: a Key to Population Estimates and Architectural Variability. In: Precolumbian Population History in the Maya Lowlands, P. Culbert and D. Rice (eds.), University of New Mexico, Albuquerque.

SEDEMA. (15 de enero de 2019). *Programa Sistemas de Captación de Agua de Lluvia en Viviendas de la Ciudad de México*. Obtenido de Secretaría de Medio Ambiente: https://sedema.cdmx.gob.mx/programas/programa/programa-de-sistemas-de-captacion-de-agua-de-lluvia-en-viviendas-de-la-ciudad-de-mexico

Sky Ha. (2018). *Sky Ha' Rainwater*. Obtenido de http://www.skyharainwater.com/proceso/

Soluciones Hidropluviales. (2019). *Hidropluviales.com.* Obtenido de Captación agua de lluvia: https://hidropluviales.com/2018/07/05/captacion-agua-de-lluvia-2/

UN. (2018). Sustainable Development Goal 6. Synthesis Report 2018 on Water and Sanitation. UN Publications, Geneva.

WWAP. (2015). The World Water Development Report 2015: Water for a sustainable World. UNESCO, Paris.

Chapter 6

Harvesting rainwater: An adaptation strategy for peace and the climate crisis

M. Pacheco

Natura Foundation chair of the board, Bogota, Colombia
Margapacheco0811@gmail.com
@margamiel

Keywords: Andean Amazonia connectivity, climate crisis, deforestation, flying rivers, new skills, rainwater exchanges

6.1 INTRODUCTION

An independent environmental thinking platform has been growing since the early 1980s in to the 20th century in Latin America. It is relevant to understand why it has taken such a long time to position rainwater harvesting as an adaptation tool to face poverty, access to fresh water, cultural diversity and climate vulnerability. In Colombia, it can also be associated with the complexity of the peace building process, a political situation that has affected deforestation after the peace agreement between FARC and the State. This particular situation has affected the Amazonia forest 'water factory', as deforestation in the year 2019 has been increasing dramatically. Rain and all its forms of utilization in the Andean Amazonian basin is produced by a hydrological cycle determined by equatorial geography and biological conditions. Ecological functions and connectivity are central to produce rain. Protection of the cradle of rain in the tropics means

 International Rainwater Catchment Systems Experiences: Towards Water Security
Editor(s): José Arturo Gleason Espíndola, César Augusto Casiano Flores, Raul Pacheco-Vega and Margarita Rosa Pacheco Montes
doi: 10.2166/9781789060584_0083

ensuring the health of 'the flying rivers', which guarantee life in the continent and in the world.

This editorial initiative, born in Mexico, has permitted the construction of a tropical network of a variety of rainwater harvesters. Since Arturo Gleason in Guadalajara organized the 2017 meeting to collect experiences and launched the Mexican Rainwater Association AMSCALL (in Spanish), IWA Publishing in London started the process of inviting participation in a 'Contributors Agreement'. This idea of assembling different points of view from all continents on a voluntary basis has led to a legacy for future generations. As the world is facing a global climate crisis, our contribution in this publication concerns a variety of local experiences in three continents with adapting socio-cultural environments.

In Colombia, located at the north of the equatorial countries in Latin America, between two oceans and with half of its territory in the Amazonia and Orinoquia basins, rainwater harvesting is still an ancestral practice, barely recognized by the Nation State as an adaptation strategy to the current climate crisis. Rainfall patterns are changing rapidly due to deforestation and fires in the forest expanding to pave the way for alternative uses. As rain depends on the heath and connectivity of ecosystems to ensure the water cycle, acting at national and international levels is necessary to guarantee action at local levels. Collection, storage, distribution, social sharing and final disposal of rainwater utilized for multiple purposes urge political action and civil society concern. Rainwater harvesting (RWH) values should become part of the circular economy principles of reuse, recycle and reduce waste and of the implementation of the Sustainable Development Goals, 2, 3, 6, 10, 11 and 13, where water is central for life. The aim is to scale up rainwater harvesting experiences and to strengthen more institutional support from national and local governments. In the education sector, this book should provide useful information for training schools and to generate new skills to improve access to rainwater. It was Augusto Angel Maya, who in the 'Challenge of Life – Ecosystem and Culture, IDEA, Bogota 1996 said, 'The human species and the culture, belong to a natural order in the same way as plants or animal species. It is the evolutionary process that conducts towards an instrumental adaptation, social organization and symbolic elaboration. With technology, these should be all included as part of nature's evolutionary process'.

6.2 CONFRONTING GLOBAL CLIMATE CRISIS

Youth worldwide is clamouring for political change to stop the global climate crisis. Current awareness is increasing, but the speed of confronting the situation to prevent water and food scarcity is at a slower pace. Climate variability in both hemispheres is being evidenced by rapidly melting glaciers, rain reduction due to deforestation and extensive cattle breeding, flooding and droughts, increasing seasonal temperatures with health impacts, extreme storm events, climate displaced population, increasing urban air pollution, and in the broader picture, social uncertainty. Media, social networks, academia and geographical information systems worldwide are communicating planetary changes.

In Andean Amazonia countries hosting the most strategic ecosystem producing fresh water in the sky and being the main CO_2 sink, the hydrological cycle is at risk from rapid deforestation and induced fires, changing land use and biodiversity functions. Global uncertainty has been expressed by the Intergovernmental Panel for Climate Change IPCC in many reports and civil society organizations and youth movements are influencing public opinion concerns.

The political position of the Brazilian government to promote 'colonization' of Indigenous Peoples territories under the argument of 'progress and sovereignty' is endangering the source of rainwater harvesting options in the Latin-American region. The threat of depleting the Amazon is provoking irreversible effects and risks a massive population displacement process and food scarcity. These phenomena in the basin are environmental services that affect rain production and increase air pollution in cities and towns in Latin America. Protecting the 'flying rivers' (Pacheco, 2018), i.e. feeding the clouds from the hot evaporation systems of the forest is at stake.

The 'flying rivers' provide rain to all Andean urban centers and rural areas. Weak institutional monitoring systems to control deforestation, combined with long lasting corruption, illegal drug trade and river gold mining, plus land occupation for pastures and post-conflict armed groups in key spots, endanger the cradle of forest–ocean–mountain water production. In Colombia, after the peace agreement signed by the FARC guerrilla and the Colombian State in December 2016, large amounts of capital and money laundering have been inducing forest fires, depleting biodiversity and introducing new farmers and extensive palm-oil monocultures, so changing water production patterns. Justice is backing up with thousands of demands with slow responses.

Amazonian departments of Caqueta, Amazonas, Putumayo, Guaviare, Guainia, Vaupes Vichada and south of Meta, 46% of Colombian territory, are a world cultural and natural patrimony heritage for water and peace preservation. In that context, rainwater harvesting analysis starts by protecting the sources of life.

6.3 RAIN IN THE ANDEAN AMAZONIA WATER CYCLE

The rain cradle in the Andean hydrological cycle has a particular feature. In tropical geography the 'Paramos' ecosystem, unique in the world, make the highlands (3,200 meters above sea level) the rain sink and a river system cradle. The Paramos ecosystems are fed by the 'flying rivers' coming from Amazonia and Orinoquia where vast plains and forest land play a regulating climate role, intimately connected with the Pacific and Caribbean wind and steam raising currents. In the highlands, water produced at the 'Paramos' by endemic species, store rain and mist like sponges and in lakes which help infiltration and runoff. Highland lakes collect and store rain, making 'Paramos ecosystem' essential. In the case of Bogota Capital District, Chingaza and Sumapaz National Parks at 3,400 meters above sea level (Figure 6.1) are the main providers for a city region of 11 million people and the surrounding region. This is understood to be the largest scale rainwater harvesting concept. The National Natural Park System and the Natural Civil Society Reserve Network have been contributing to protect highland water sources jointly with regional environmental authorities, as part of the National Environmental System, SINA (in Spanish).

SINA celebrated in 2019 twenty-five years of existence, created after the UN Earth Summit in Rio 1992, by Law 99 in 1993. Today, SINA is weakened by political interests controlling investments and contracts for large oil and mining extractive sector projects, dams and infrastructure, affecting forest conservation. Main rivers are under threat by cities and towns with non-existent sewage treatment plants, thus receiving domestic and industrial polluting waste. In this context, RWH and wells are healthier than using water from rivers (polluted with mercury) so fishing is endangered for human consumption.

Figure 6.1 Sumapaz tropical highland ecosystem absorbs the "flying rivers" blown from Amazonia, where deforestation is affecting the rain patterns and increasing the climate crisis. Bogota Capital District, 2019. (*Source*: Author's own).

In spite of significant international cooperation contributing to restoring peace and local economies in Amazonian indigenous communities, uncontrolled and illegal deforestation in different regions of the basin contributes to increasing the climate crisis at national and global levels. The highlands, the lowlands, Orinoquia and Amazonia forests and the two oceans of Colombian tropical geography have an intimate connectivity and interdependence to produce sky water. These are the 'flying rivers' conceptual framework, described by indigenous cosmology and knowledge of the functioning of their habitat.

Effects due to deforestation and fires are transforming cloud formation and rain in Andean cities. Pacific Ocean temperatures influence wind and transit of the flying rivers in every tropical floor of the Andes, and are impacting rainfall patterns. 'El Niño' and 'La Niña' weather events are becoming more intense and more frequent (Euscategui & Huetado, 2011), the worst being registered in 2010–2011.

Nation States are timidly recognizing the multi-purpose role of rain to face the global climate crisis. In Colombia, water management and forest conservation and restoration have different institutional responsible managers both at national and territorial levels. Water management separation puts rain in a shadow place, where harvesting rainwater is not a key issue in national and local policy and in urban environmental regulations. RWH is left to private voluntary initiatives with no significant relevance in environmental planning.

Water and Sewage Services at municipal level sell water but do not promote or stimulate rainwater harvesting systems as water is considered a commodity and not a public good to be implemented as an adaptation strategy. Scattered urban experiences are mainly private initiatives interested in promoting principles of circular economy, but in rural areas, RWH experiences respond mainly to family survival needs and systems are built with no sponsorships, technical assistance or subsidies.

At an international basin level, the forty-year old Amazon Cooperation Treaty Organization (OTCA in Spanish), is now superceded by critical climate changes occurring in the eight countries within the basin. The OTCA headquarters in Brasilia has been silent on the fires, rainforest destruction, cattle breeding and extensive monocultures changing indigenous communities ancestral tenure and wise biodiversity conservation. The Andean Amazonia connectivity is been transformed by political decisions at multinational level by tolerating new forms of colonization. Tropical rain forest is devastated with a risky and irresponsible idea of 'progress', which has no long-term vision.

6.3.1 Scaling rainwater harvesting

Strategic ecosystems connecting the water cycle form the complex hydrological system of a sky–sea–soil–mountain circular motion system pushed by the winds (Figure 6.2). In this period of high climate vulnerability, when a growing young citizen movement is advocating decarbonizing society, education and media play

Figure 6.2 Scaling up RWH means understanding tropical ecosystem connectivity. Serrania La Lindosa, Guaviare, Amazonia, 2019. (*Source*: Author's own).

a key role in the scale up of rainwater utilization. The question is how to confront the effects during the current geological era, the Anthropocene (Kolbert, 2015). Human activity is affecting the weather and food security, public health, and behavioral changes are appearing in all living species.

Research findings are responding to resilience limits, extensive lists of scientific publications are warning the international community. Leadership from thematic networks and from concerned media is being raised. Voices from non-state actors are claiming concrete action. These are urging the strengthening of institutional arrangements to cope with the national and local energy transition process, improve adaptation and mitigation strategies, and increase technical training to scale up circular economy, in particular within the water sector.

To scale up rainwater adaptation strategies, environmental land use planning should include resilience and adaptation plans, subsidies, vocational training and dissemination. The global goal suggested by scientists from IPCC in Korea in 2018 should justify the strengthening of public education and improve multipurpose rainwater utilization in urban and rural areas in every municipality in the tropics, associated with the energy transition process.

As a public good, rain collection, storage and use, should be free of charge. In Colombia, island communities in San Andrew and Old Providence Islands are using their rain storage as part of the house structure. All schools and houses in Amazonia have their own local systems and this is also true in many rainy regions, including the Choco region, the rainiest, richest in biodiversity and poorest afro and indigenous population in Colombia. Small municipalities with weak or inexistent public water and sewage services lack regulations to encourage rainwater utilization, not recognizing the cultural traditions which adapt to local environments.

Multipurpose rainwater options with potential uses for household and productive uses, such as sanitation and hygiene, watershed management, food sovereignty and soil conservation, aquifer recharge and ecosystem conservation, runoff management, disaster prevention and reconstruction, should be incorporated in to integrated water resource management policies (IRHA, 2006). All RWH scales are valid to give an added value to rainwater utilization.

The climate change crisis should thus encourage scaling up of RWH at policy level, permitting urban and rural housing and public infrastructure to use rain as a public good in non-potable water uses: washing and cleaning options, gardening, landscaping, animal feeding and biodiversity conservation, amongst others. Legislation to scale up rainwater utilization should also cover different sectors, from housing, health, energy, education and agriculture, to larger scales in land use and environmental planning and final disposal.

Building resilience to confront extreme weather events, as storms are becoming more frequent and more intense, should mean designing institutional capacities and arrangements within sectors to facilitate citizen's adaptation to climate change.

6.4 KNOWLEDGE EXCHANGES

In the call that the Intergovernmental Panel for Climate Change (IPCC) made to cities in 2018 to include adaptation strategies, different territorial levels should be included. From the basin to the neighborhood and house level, integrated RWH approaches should be developed to connect experiences, promote benefits and prevent difficulties, encouraging scaling up and knowledge sharing and exchanges among cities and regions.

There is a citizen's 'de facto' knowledge on rain utilization, made up from local knowledge on weather and seasonal agricultural cycles in rural inland, highlands and coastal areas. This knowledge belongs to a precautionary principle responding to satisfy human needs. An example is the amphibian culture (Figure 6.3) in wetland regions in the Colombian Caribbean, where Senu and Chimila indigenous traditional knowledge on flooding control in the Magdalena River marshes area dealt technically with issues of flooding by building polders.

Figure 6.3 Amphibian houses adapted to raifall and flooding in Putumayo, Amazonia. (*Source*: Author's own).

This tradition has unfortunately disappeared since Spanish colonization and the introduction of cattle breeding in wetland regions.

Indigenous conservation knowledge in the Amazon basin and in the highland forest should be better disseminated, particularly to learn from it how ecosystem conservation has been implemented to protect survival; forests in the lowlands, valleys and highlands have an intimate connectivity to maintain the water cycle and the sacredness of rain, wind, air and soil, essential to spiritual beliefs connected with nature.

At the city level, new urbanizations and housing projects should learn from other cities by involving RWH systems in urban regulations. From roof and wall rain collection to lakes in green spaces and run off utilization for gardening and infiltration in parking lots, economical and ecological benefits should become urban adaptation regulations to confront the current climate crisis.

6.5 INTEGRATED RAINWATER MANAGEMENT APPROACH

An Integrated Rainwater Resources Management (IWRM) approach links social and economic development with the protection of vital ecosystems. The aim is to increase the multipurpose use of rain and reduce potable water consumption to strict needs. The rationality today is the payment to access water, waste and sewage services, with no intention in reducing the final volume disposed in the combined sewage system. Zero waste goals with reuse and recycling water resources need to be part of education and public awareness. IWRM promoted by national policy in Colombia requires adjustment of its goals and strategies at sub-regional and municipal levels, aiming at reducing the polluted water thrown directly to the rivers and subsequent CO_2 emissions, separating rainwater waste from domestic sewage systems. Decentralized rainwater utilization should generate a new culture to protect the 'flying rivers' and its final disposal once has been used and reused.

The complex interaction between soils, air, sea water, underground water, winds, mountains and forests in the tropical geography of Andean Amazonian countries demands specific policies to protect biodiversity and quality of life, and to regulate climate change. It is an integrated life system where rain links all. New policies to position rainwater harvesting should include a portfolio of current private and public initiatives at urban and rural levels, and understand they are all part of an adaptation strategy to consolidate peace and reduce the effects of the current climate crisis.

ACKNOWLEDGEMENT

Special thanks to Arturo Gleason for pushing this editorial project forward. Also to Rachel Nisbet from the International Rainwater Harvesting Alliance IRHA in Geneva, for her translation of Gilbert Sene's text from French into English. We hope that this IWA Publishing will serve as a pedagogical tool in rural and urban English speaking communities, disseminating and sharing knowledge from Spanish and French speaking regions and communities in the tropics.

REFERENCES

Euscategui C. and Huetado G. (2011). *Análisis del impacto del fenómeno "La Niña" 2010–2011 en la Hidroclimatologia del país*, Bogota: IDAEM.

IRHA (2006). Integrated Rainwater Resource Management. Promotional leaflet. Geneva: IRHA.

Kolbert E. (2015). *La Sexta Exinticion*. Critica, Barcelona, Spain.

Pacheco M. (16 de 08 de 2018). *Estrategia adaptativa: aprovechar el agua de lluvia*, Obtenido de https://www.semana.com/opinion/articulo/los-beneficios-de-aprovechar-el-agua-de-lluvia-columna-de-margarita-pacheco/579753.

Section 3

Programs

Chapter 7

Social enterprise on sky water harvesting for solving drinking water crisis in coastal areas in Bangladesh

M. Murase, PhD[1,2]

[1]*Chairman, Sky Water Bangladesh Ltd.*

[2]*CEO, Institute for Sky Water Harvesting Ltd.*

Keywords: arsenic contamination, diarrhea, salinity, Sky Water Harvesting, social enterprise

7.1 DRINKING WATER CRISIS IN COASTAL AREAS OF BANGLADESH

At the moment, there are still more than 20 million people in rural areas of Bangladesh who have no access to safe drinking water. Women and children must take more than one hour every day to fetch water from the pond. Sometimes the pond is located 4 km from their house; water fetching is a very heavy burden. Local people sometimes buy water from a water seller so they don't have to fetch water themselves, but it costs 6 Tk (1USD = 80 Bangladesh Taka) per 20-liter barrel (NEC Corporation, 2013). This is an additional economic burden. Moreover, the water source for local people and water sellers is the same pond which is polluted by water-borne pathogenic microorganisms. Countless people who have drunk pond water suffered from severe diarrhea, in some cases causing death. It has been estimated over 45,000 under-five children die every year in Bangladesh from diarrhea caused by contaminated water (WHO, UN-Water, 2017).

Besides, to the diarrhea issue, more than 35 million people are under threat of arsenic contamination from groundwater in 59 of 64 districts. There are currently

Editor(s): José Arturo Gleason Espíndola, César Augusto Casiano Flores, Raul Pacheco-Vega and Margarita Rosa Pacheco Montes
doi: 10.2166/9781789060584_0095

1.4 million tube wells painted red which means they are contaminated with arsenic at a level which exceeds national drinking water standards (0.05 mg/l). Arsenic contamination is not artificial pollution; it originated from natural soil contamination.

Some local people continue to drink arsenic-contaminated groundwater. They can not recognize arsenic poison as it has no taste, no smell, and no color. Arsenic is easily absorbed and accumulated into the human body. After drinking arsenic-contaminated water over a long time, people suffer from melanosis, symmetric hyperkeratosis, and skin cancer in the worst case (Islam & Islam, 2018). Also, groundwater and pond water in the coastal areas have been affected by saline damage, which has been accelerating year by year due to the rising seawater level caused by climate change and shrimp farming. It is difficult to remove salinity with the pond sand filters installed for purification of pond water. In addition to the saline issue, almost all tube wells are heavily contaminated with iron. Local people have been suffering from bad tasting drinking water and food which is cooked in this water.

In 2011, in cooperation with JICA and local NGOs, I started a base of the pyramid (BOP) social enterprise for sky water harvesting as CEO of the Institute for Sky Water Harvesting Ltd (ISWH) to solve the drinking water crisis in rural areas in Bangladesh. The basic concept of social enterprise is the effective use of available materials and human resources in local societies. That is, for example, in the AMAMIZU project described in this chapter, materials for an AMAMIZU tank, such as mud, sand, and cement should be purchased in local areas, and AMAMIZU masons should be nurtured through job training of local people. Also, transportation of the AMAMIZU facilities to customers should be carried out through local transportation methods such as rikisha vans, engine vans, and boats. Social projects make the local economy sustainable.

7.2 BASE LINE SURVEY

ISWH started a baseline survey regarding a BOP social enterprise on sky water harvesting in cooperation with JICA and local NGOs at coastal areas in Bangladesh in 2011. 300 local people were surveyed on drinking water issues, awareness of sky water harvesting, affordable price of sky water harvesting, etc. The result was very interesting. Almost all local people replied that pond and groundwater have an odor and are discolored and also that fetching water is a heavy burden for them. And they recognized that diarrhea is caused by water contamination (Table 7.1 & Table 7.2) and that sky water harvesting is a key solution to solve the diarrhea issue.

More than 50% of local people could afford up to 3000 Tk for a tank. Regarding this, we had one more important piece of evidence for social enterprise. Local people have spent on average 1425 Tk for medical expenses to treat water-borne diseases and 1416 Tk for buying water, including transport of water (Figure 7.1 & 7.2). It means if local people buy a tank for 3000 Tk, they could reduce their overall

Table 7.1 Baseline survey of drinking water issue in Morrelganji (1).

	Village-1 Baroikhali	**Village-2 Tetubari**	**Village-3 Kalikabari**	**Village-4 Bashbaria**	**Total (n)**	**All Villages %**
What do you think 'Water is the main reason of health problems here'?						
Not at all	4%	8.3%	0%	0%	3	3.1%
Some extent	12%	8.3%	13%	16%	12	12.4%
Definitely	84%	83.3%	87%	84%	82	84.5%
Total	100%	100%	100%	100%	97	100%
Your friend's Opinion about Drinking waterproblem (What do you think about your friends, neighbors opinions or judgement on drinking water problem of the locality)						
Good	0%	0%	0%	0%	0	34.3%
Okay (Not bad)	4%	0%	4%	0%	2	62.6%
Bad	96%	100%	96%	100%	98	3%
Total	100%	100%	100%	100%	100	100%

Table 7.2 Baseline survey of drinking water issue in Morrelganji (2).

	Village-1 Baroikhali	Village-2 Tetubari	Village-3 Kalikabari	Village-4 Bashbaria	Total (n)	All Villages %
Is there 'order' in your drinking water?						
No Order	8%	0%	4%	0%	3	3%
Little Order	68%	66%	56%	56%	59	59%
Strong Order	24%	44%	40%	44%	38	38%
Total	100%	100%	100%	100%	100	100%
Is there 'Color' in your drinking water?						
Very dirty and Muddy	20%	33.3%	44%	44%	34	34.3%
Not so clean	68%	66.7%	56%	66%	62	62.6%
Clean	12%	0%	0%	0%	3	3%
Total	100%	100%	100%	100%	99	100%
Distance and time; Do you think fetching water is burdensome for you in terms of distance and time?						
A great deal	92%	100%	91.3%	100%	88	34.3%
Some deal	4%	0%	8.7%	0%	3	62.6%
Not at all	4%	0%	0%	0%	1	3%
Total	100%	100%	100%	100%	92	100%

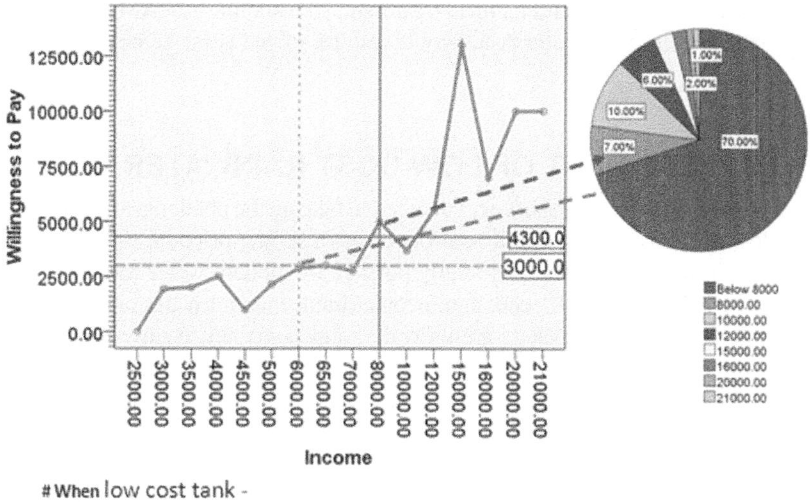

Figure 7.1 Baseline survey of affordable cost of rainwater jar. (*Source*: Author's own).

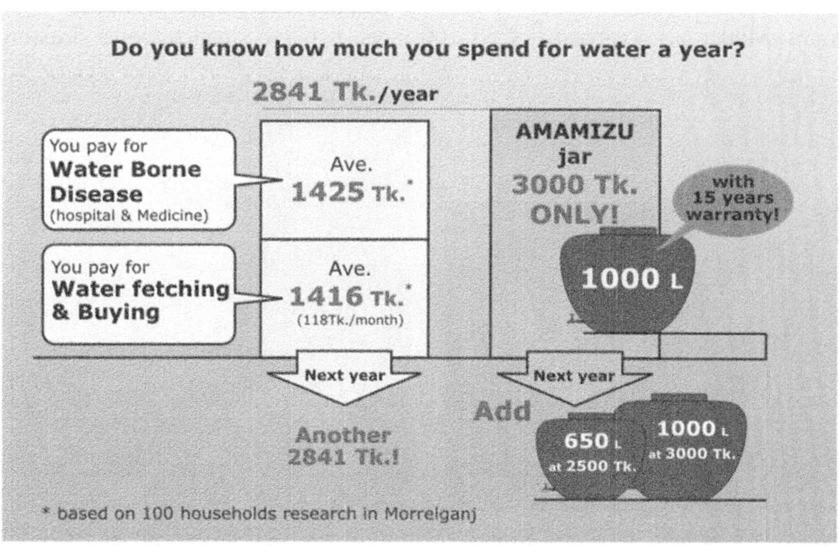

Figuer 7.2 Baseline survey of water cost. (*Source*: Author's own).

household costs for water and medical treatment, so installing AMAMIZU is not so expensive. The drinking water issue would also be solved at the same time (WHO, UN-Water, 2017).

7.3 DEVELOPMENT OF LOW-COST RAINWATER TANK

Based on the above baseline survey, I decided to take up the challenge of developing a new low-cost tank which could be produced at a selling price of 3000 Tk.

Before drinking water was artificially purified by modern water supply systems in Japan, which started 128 years ago, people drank 'natural water', such as, river water, lake water, spring water, groundwater, and rainwater. Rainwater has been the main source of drinking water in isolated islands in Japan. Bangladesh has the same history. Local people in rural areas collected rainwater to drink in an earthenware pot called a 'Motka' (Figure 7.3). Still now they harvest rain into a Motka and use it for drinking and cooking in the rainy seasons.

When I went to a customer's house to install a concrete rainwater tank six years ago, I discovered several Motka in a line filled with rainwater at the back of the house. I asked the house owner why they did not drink rainwater. They replied, 'Yes, rainwater is safer than pond water, that's why we have reserved rainwater for our friends and relatives'. After 10 minutes, when rainfall with high intensity flushed out dust and leaves on the roof, local people harvested clean rain through galvanized gutters into a Motka with its mouth covered with cloth. The cloth filters the water and prevents invasion from mosquitos. Over time local people have learned how to harvest safe rainwater through their own experiences.

Motka is cheap and has been affordable for poor local people. But it has weak points. Motka has a capacity of 50–100 liters. It is too small to cover drinking

Figure 7.3 Motka. (*Source*: Author's own).

Figure 7.4 Thailand giant jar. (*Source*: Author's own).

water needs over the dry season and is also fragile. I came up with the idea of new jumbo Motka which is solid and has a larger capacity than traditional Motka.

To meet the above requirements, I found a mortar or concrete water jar in Thailand, called 'Giant Jar' (Figure 7.4), which has been used by local people in the northwest rural areas. The capacity of this jar was between 650 and 2000 liters. The selling price of a 650-liter Giant Jar was 3000 Tk. Thais have harvested sky water into the Giant Jar to use as safe drinking water. They purchase an additional Giant Jar every year. There were 4–6 Giant Jars in a house.

ISWH dispatched two Bangladeshi masons to the National Chombri Job Training Center in Thailand to acquire the skill of Giant Jar technology. After training, they came back to Bangladesh. ISWH hired them and started to develop the Bangladesh version of a low-cost water container. This technology is popular in Thailand and Cambodia, but not found in Bangladesh. ISWH improved the Giant Jar technology.

The production method is as follows: (1) make a mother mold; (2) make a dice based on the mold, a die consists of 16 parts; (3) third, set up the dice on the base and fix with wire (Figure 7.5); (4) mud treatment of the outside wall followed by mortar treatment on the mud layer (Figure 7.6); (5) reinforcement with steel wire net and mortar (Figure 7.7); (6) remove the dice from inside the jar after the mortar has dried (Figure 7.8). It looks a little like an apple which has been hollowed out in the middle. All die can be reused.

ISWH could complete the Bangladesh version of the Giant Jar, which is made from mortar reinforced with steel wire net and has a capacity of 1000 liters. This new low-cost jar was named 'AMAMIZU'. AMAMIZU in Japanese means 'sky water'. I decided to call rainwater 'sky water' through learning the wisdom of our ancestors. Sky water is a gift from heaven. Our ancestors have harvested rain and used it for drinking water and they called rainwater 'sky water' with great respect. AMAMIZU also has another meaning. It means 'sweet water'. When Bangladeshis drank sky water instead of pond water, which has high salinity, they told me that sky water tasted sweet.

Figure 7.5 Set up dice. (*Source*: Author's own).

Figure 7.6 Mortar treatment on the mud layer. (*Source*: Author's own).

Figure 7.7 Reinforcement with wire net. (*Source*: Author's own).

Figure 7.8 Remove dice from inside the jar. (*Source*: Author's own).

ISWH obtained the patent for AMAMIZU systems in February 2015. A description of the AMAMIZU systems is given below and in Figure 7.9:

- *Catchment:* The sky water catchment area should be the roof. Roofs made of anything except jute are acceptable, for example, tin, concrete and mortar or ceramic tiles.
- *Collection*: Rain falling on the roof is collected by a plastic gutter and flows into a jar through chains hanging from the gutter. Chain gutter systems have been used commonly at temples in Japan. But we experienced problems with the chain gutter system: whenever we had heavy rain, water splashed from the plastic chain into the house through open windows. Now we use a movable elbow pipe instead of a plastic chain.
- *Storage*: Sky water is stocked in AMAMIZU. When an AMAMIZU tank fills up with rain, it flows to the next AMAMIZU through an overflow pipe.

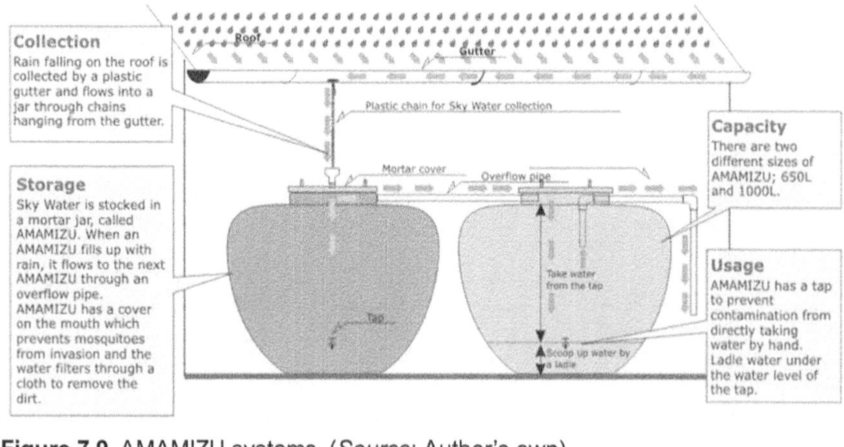

Figure 7.9 AMAMIZU systems. (*Source*: Author's own).

AMAMIZU has a cover net on the inlet pipe which prevents mosquito invasion. Sometimes local people put a cover on the mouth of AMAMIZU and filter the water through a cloth to remove the dirt. One AMAMIZU can connect with another one through a connection pipe.

- *Usage*: AMAMIZU has a tap to prevent contamination from directly taking water by hand. The tap is made from plastic, but metal taps are also available which can be locked.
- *Transportation*: One AMAMIZU was transported by a Rikisha van to customers living within 2 km. Engine van instead of Rikisha van is available for transportation to customers more than 2 km away. But we had trouble. The total weight of AMAMIZU is 250 kg. The AMAMIZU body was affected by strong shaking during transportation which caused crack issues. From this lesson, we have developed a new AMAMIZU, called 'Super AMAMIZU' which was reinforced with steel wire net by the adaptation of ferro-cement technology. Also, we have transported AMAMIZU to stock yards by an engine boat which are located more than 4 km distant. Nine AMAMIZU could be transported this way at one time.

The AMAMIZU production center (APC) was constructed in Morrelganji in 2012 in cooperation with JICA.

7.4 AMAMIZU SOCIAL PROJECT

ISWH started a pilot project of AMAMIZU social enterprise in cooperation with JICA and PR Bangladesh (a local NGO which I established in 2008) in 2012. The price of AMAMIZU, including installation and gutter pipe was 3000 TK and transportation was 1300 TK. The total selling price of AMAMIZU systems was 4300 TK. So that more BOP (Base of Pyramid) people could buy AMAMIZU systems, we introduced a more flexible payment method. After local people paid the down payment of 2000 TK, AMAMIZU systems were installed. PR Bangladesh raised awareness of sky water harvesting (SWH) and provided operation and maintenance training of AMAMIZU systems to local people. The remainder was reimbursed within 6 months. The pilot project was successful, 200 AMAMIZU were sold and installed. 97% of remaining payments were recovered.

Following the pilot project, Skywater Bangladesh Ltd. (SBL) was established to tackle the social enterprise side of sky water harvesting. A production manual of AMAMIZU was prepared for quality control. Patent of AMAMIZU systems and design registration of AMAMIZU were secured. Also, a training center was opened to develop skilled masons and water quality control. A newsletter published the results of monitoring and was also used for marketing and awareness. The total number of AMAMIZU produced reached 3600 in March 2018.

Figure 7.10 CHB tank of Morrelganji Health Complex. (*Source*: Author's own).

The main target of the AMAMIZU social project is private houses. But for supply of safe drinking water for all, the social enterprise of sky water harvesting should be promoted not only at private houses but also at community and public facilities in rural areas where there are no piped water supply systems. Application of the sky water harvesting technologies to community and public facilities should be different from AMAMIZU systems. SBL developed the concrete hollow block tank (CHB tank) in 2013. Capacity could be scaled up from 5 tons to several hundred tons. In September 2013, in cooperation with JICA, SBL constructed CHB tanks with a total capacity of 150 tons (50 tons × 3 unit) at Morrelganji Health Complex, Bagerhat district, to supply safe drinking water to patients and medical and official staff (Figure 7.10).

Before installation of sky water harvesting systems, water for drinking and cooking was taken from a pond in the hospital. But when there were water shortages in the dry seasons, river water which flows nearby the hospital and is contaminated with domestic waste water and salinity was drawn into the pond. A sand filter had been installed for purification of the pond water, but it is difficult to remove pathogenic microorganisms and salinity. Many people suffered from diarrhea.

A Health Complex management committee for the maintenance of sky water harvesting systems has been organized with medical doctors and official staff members in cooperation with SBL. They clean the roof catchment areas every month and made rules about intake of sky water from the tank so as to secure drinking water for the whole the year. The hospital has been satisfied with the quality of their drinking water by good ownership.

In addition to Morrelganji Health Complex, in 2019, seven CHB tanks whose capacity is 50 ton have been installed to secure safe drinking water in an emergency at seven cyclone shelters in Chittagong City where a cyclone resulted in a drinking water crisis.

Besides coastal areas, such as Morreganji, there are many places experiencing the same drinking water crisis, such as in the northern areas of Bangladesh and the hill tracts of Chittagong. There are more than 20 million people who have no access to a safe drinking water source. This means there is a big demand for sky water harvesting. A sustainable social enterprise model of sky water harvesting is shown in Figure 7.11. SBL has three social enterprise plans for different targets. The first is AMAMIZU social enterprise for BOP people, called the mother business, which is production, sales, and installation of AMAMIZU (1000 liter). The second is high-quality concrete ring (CR) tanks (4400 liters) for people with a higher income. The third is concrete hollow block (CHB) tanks (more than 5000 liters) for communities and public facilities. AMAMIZU business is low profitability compared to CR and CHB tanks. The profit made on CR and CHB tanks is used effectively for the mother business and mason training. Also, donations are used for making the mother business sustainable.

SBL will start social franchise (NORENWAKE) systems to promote sky water harvesting all around Bangladesh in the future. NORENWAKE is part of Japanese traditional commercial culture, which has secured a high-quality product and sustainable service with ownership. After 10 years of experiencing the sky water harvesting social enterprise, if some skilled masons and staff want to take

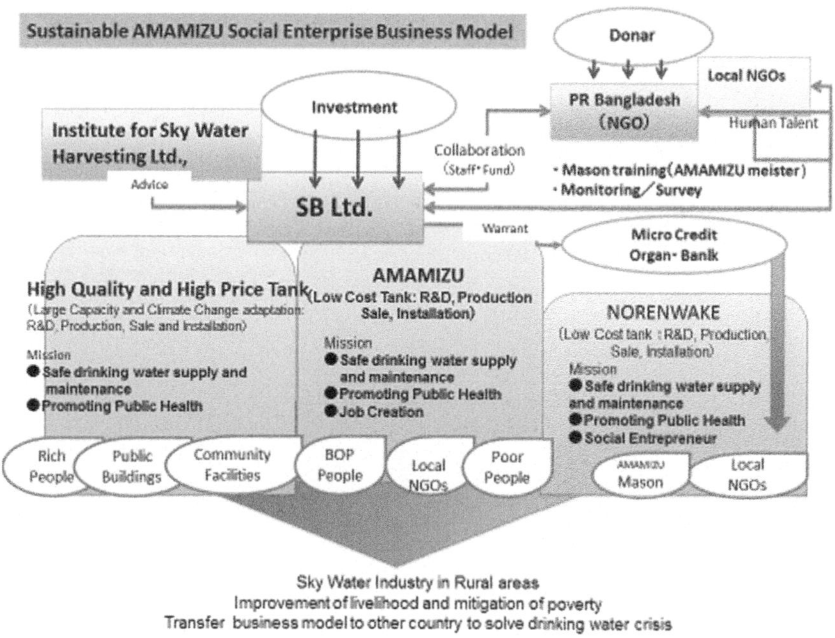

Figure 7.11 Sustainable AMAMIZU social enterprise business model. (*Source*: Author's own).

up the challenge of NORENWAKE, they could become independent with microcredit systems and start to produce, sell and install AMAMIZU in cooperation with SBL. If AMAMIZU NORENWAKE could spread all around Bangladesh, it could make a significant contribution to solving the drinking water crisis in rural areas of Bangladesh. We hope it will create a 'sky water industry' by circulating human talents, materials, and sustainable funds in local communities and become part of the social fabric.

REFERENCES

Islam M. and Islam F. (2018). Arsenic Contamination in Groundwater in Bangladesh: An Environmental and Social Disaster. IWA Publishing, London, UK.

NEC Corporation (2013). Preparatory Survey on BOP Business on the Smart Village in India, Final Report Summary, The Institute of Cultural Affairs Japan, India.

WHO/UN-Water (2017). UN-Water Global Analysis and Assessment of Sanitation and Drinking-Water (GLAAS) 2017 Report: Financing Universal Water, Sanitation and Hygiene under the Sustainable Development Goals. UN Water, Geneva, Switzerland.

Chapter 8

Catalyzing the widespread adoption of rainwater harvesting in Mexico city

D. M. Vargas[1] and E. Lomnitz[2]

[1]Isla Urbana, President, Épsilon 130, Colonia Romero de Terreros, Delegación Coyoacán, CDMX, CP 04310, México
[2]Isla Urbana, Director, Épsilon 130, Colonia Romero de Terreros, Delegación Coyoacán, CDMX, CP 04310, México

Keywords: decentralized water management, rainwater harvesting, scalable water solutions, social entrepreneurship, urban sustainability

8.1 INTRODUCTION

The Isla Urbana (IU) project seeks to 'detonate' adoption of rainwater harvesting (RWH) systems by Mexican households as a response to the water crisis. By installing these systems in households, a sustainable water source is secured that directly benefits the families involved. When greater numbers of houses begin harvesting rainwater, their collective reduction in demand and greatly increased self-sufficiency builds resilience into the community and the city as it faces shortages from conventional sources.

IU designs systems that take advantage of certain idiosyncratic characteristics of Mexico City's houses that allows for very effective RWH at low costs. It has been based since its founding in a low income, water scarce area of the city and has developed much of its work there. Constant close contact with the systems and their users has provided a living research and development lab where ideas and

Editor(s): José Arturo Gleason Espíndola, César Augusto Casiano Flores, Raul Pacheco-Vega and Margarita Rosa Pacheco Montes
doi: 10.2166/9781789060584_0107

Figure 8.1 Isla Urbana's impacts in numbers from 2009–2018. (*Source*: Isla Urbana, 2018).

designs are tested, adjusted, retested, trashed or adopted based on their effectiveness in real people's lives.

IU has installed over 10,000 RWH systems in its nine years of existence (See Figure 8.1). Over the next six years, they hope to install another 100,000 systems in homes, benefiting over 700,000 people. The long-term goal, however, is the implementation of systems in most households in the city, which would benefit the entire population of 22 million people

8.2 SOCIAL AND ECOLOGICAL CHALLENGE

Mexico City was founded on a small island in the middle of lake Texcoco around 700 years ago. When Hernan Cortez first saw it, the city appeared beautifully integrated into its watery context. Man-made islands ringed the solid ground, crisscrossed with canals along which canoes carried goods from throughout the valley. Long causeways linked it to the shores, aqueducts brought drinking water from springs in Chapultepec, and the huge dyke of Nezahualcoyotl separated the brackish waters in the east from the fresh waters in the west and held back surges in the rainy season. Much like in Venice, Mexico's inhabitants had learned to thrive in the water, and the city seemed inextricably linked to the lake it was born in.

The war with the Spanish marked the end of Aztec civilization, and with it also began a long, ongoing battle against the lake itself. The deep relationship Mexico City had with the water around it would be thoroughly destroyed. During the war, the aqueducts and dyke were wrecked to deny the besieged city clean water, and to flood out its struggling resistance. After it's palaces and temples had been dismantled and replaced with the churches and buildings of the New Spain, the city suffered chronic floods and epidemics. The Vice-Royal government struggled to establish its capital in the swamp and considered relocating to a more salubrious location. But finally, in 1607, a different course was chosen, and fateful work began on the first drainage canal, which cut an exit from the closed watershed of the Valley of Mexico and allowed the water to flow out and into the river Tula and the gulf.

With this huge project, the future course of water management in the valley was set. Instead of finding a way to live with the abundant water, as the Aztec had done, Mexico would strive to rid itself of it. From then on draining out the rains would remain an ongoing endeavor right up until the present day, in which all that's left of the once great lakes are a few scattered and shrinking ponds.

Unsustainable water management is a critical problem facing Mexico City as it develops in the 21st century. The mega city's water supply depends primarily on an intensely overexploited aquifer below the city, and secondarily on infrastructure that pumps water over 200 km from two rivers far below the city (see Figure 8.2). Overexploitation of the aquifer is causing the city to sink which cracks the network of pipes that distribute water, meaning over 30% of the city's water is then lost to leaks. Even with this intense extraction, 750,000 people in the valley of Mexico City lack access to water while throughout the country this number balloons to 18 million (See Figure 8.3). This forces urban families to buy expensive trucked water while rural families travel to collect water from impure sources. The future promises greater scarcity as predictions point to increasing demand and decreasing supply which will disproportionately affect the poor.

Mexico City's water problems are compounded by geography (it is over 2200 m above sea level and far from any coast or large body of water) and a history of poor governance. Sustainability will require watershed level management that makes the most of naturally available water resources, while resilience must involve greater local control and self-reliance. Rainwater harvesting is a tool that works towards both ends. Currently, rain runs off into a combined sewer that expels it from the valley, while the population depends on water pumped uphill from watersheds outside. The gap between demand and supply is growing rapidly, resulting in less water availability throughout the city, but affecting poorer areas disproportionately. Introducing and teaching rainwater harvesting in these areas makes use of an abundant water source that is now almost entirely

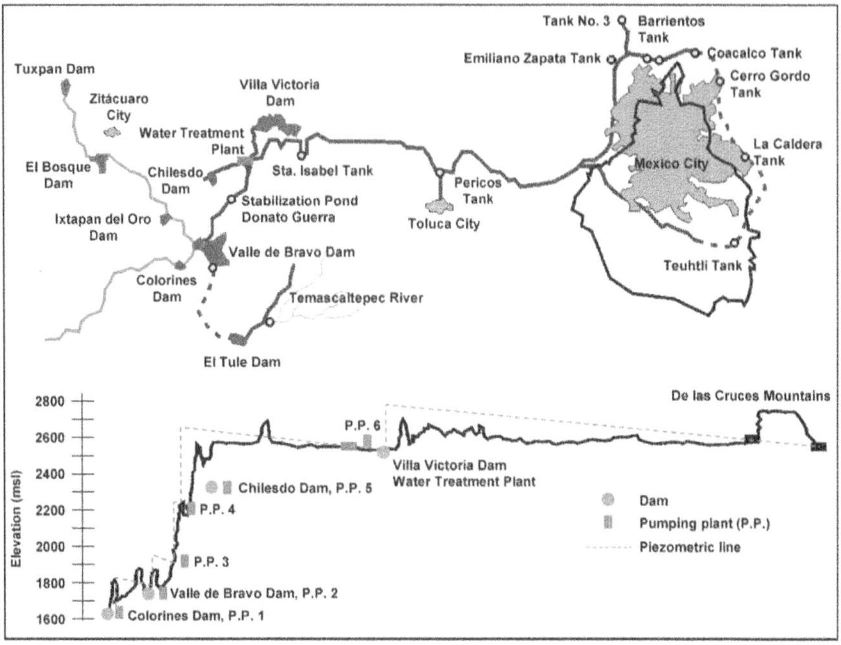

Figure 8.2 Overview of the Cutzemala Water system that provides 30% of the city's water (Tortajada, 2006).

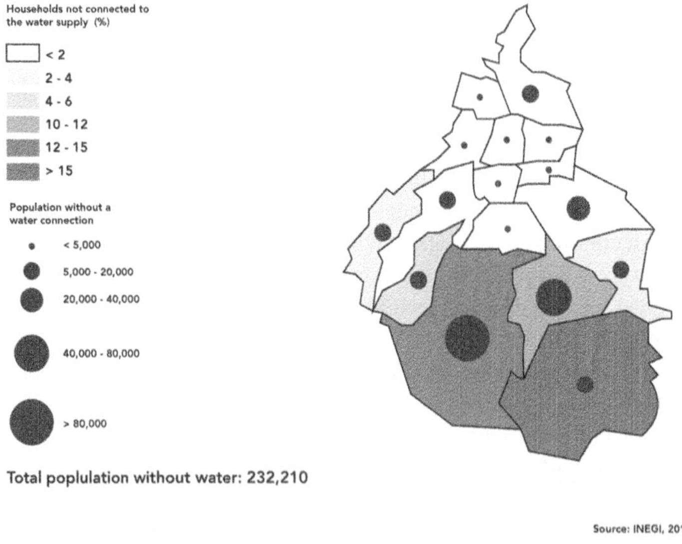

Figure 8.3 Population without access to water connection by delegation in Mexico City. (*Source*: Isla Urbana, 2015).

wasted, while reducing the population's dependence on a struggling system over which they have no control.

The overall situation is critical. With a rapidly deteriorating aquifer providing most of the city's water, and a lack of viable alternative sources anywhere nearby, projections for the future are extremely worrying. The World Bank, in association with the National Water Commission (CONAGUA) released projections calculating that, at current trends, only 50% of the city's demand could be met by sustainable sources in 2030, and at least 27% (Banco Mundial, CONAGUA, 2013) assuming the aquifer has not been depleted and can still be overexploited at current rates) will have to come from new, as-yet undefined sources.

And still, draining the rains from the Valley of Mexico continues to be a central part of the city's water management strategy. Currently, the rain that falls on the city makes up around 70% of the combined sewage that is expelled into the river Tula. Even as securing water for human uses becomes an increasingly critical problem, the floods that have affected the city since colonial times continue to devastate low-lying areas every year. The canals and tunnel systems built to drain the valley overflow during downpours, and the mixed sewage temporarily covers entire neighbourhoods. A billion-dollar drainage tunnel is currently under construction, the latest stage in the eternal struggle to rid the city of its rainwater.

Mexico City, in its evolution from a valley of lakes to a megalopolis looking anxiously at a future of thirst, presents an amazing case study in unsustainable management of water. But in recent years, an abundance of new ideas and proposals has been arising, mainly from academic and civil society organizations, which seek to fundamentally reimagine the relationship we have with water. These diverse proposals, which include ideas like regenerating Lake Texcoco in the largely empty and barren eastern section of the valley, and disentangling rivers from the sewage system and making them open once again (as was done in Seoul), all seek to plot a new course that breaks from our present trajectory, in the hope that Mexico City can once again become a place where people and water exist in harmony.

8.3 ISLA URBANA MODEL

Isla Urbana has adapted rainwater harvesting to the Mexican context to maximize impact. Most of the work focuses on low-income areas where most homes already have a cistern built to cope with constant water scarcity problems. We have designed systems that connect existing roofs and cisterns, eliminating the most expensive component one would normally need to provide. Skilled installers can put up a customized rainwater system for $1000 USD, providing between 5 and 12 months of full water autonomy to its users. The systems are designed to be expandable, and families can add storage capacity over time,

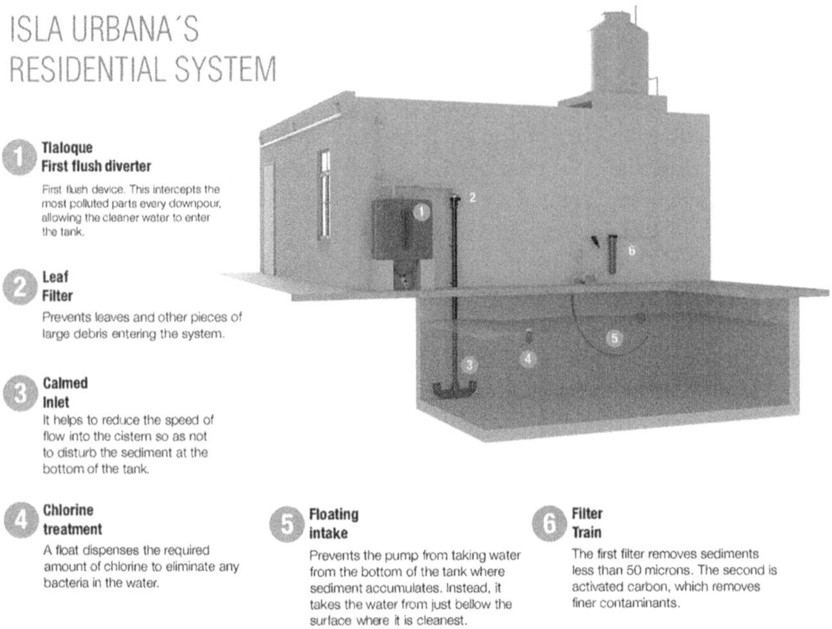

Figure 8.4 Typical RWH system as designed by Isla Urbana. The components as installed by Isla Urbana are numbered 1–6 and the more general components labelled. (*Source*: Isla Urbana).

increasing the number of months of water autonomy. As the water crisis worsens, the systems' capacity can be easily increased in response. This takes advantage of the fact that low-income families here use a gradual approach to construction anyway in order to get around the high cost of building a home. The systems are therefore dynamic and able to respond to future needs.

8.4 OTHER TESTED SOLUTIONS

The government solution to the current water shortage focuses on both the supply and demand yet still falls short. On the supply side, the government has been working to fix leaks in city pipes which has the potential to save up to 30% of the city's supply which is currently being lost. However, the plan spans 25 years and new leaks are sure to appear during that period. On the demand side, the government is implementing educational programs aimed at reducing water use among citizens. This is an important and necessary component, but changing the culture of water use and individual habits will take years. For the many areas that are not connected to the city grid, municipalities sometimes subsidize water trucks that bring water to households at a high cost. In rural areas, standard supply approaches aren't feasible because of the remoteness and subsequent high cost so the government is unable to currently provide water to villages of less than 3000 people.

So, as the future projections show, what is really needed in Mexico is another water source that is sustainable and affordable for the many low-income families. Current proposals for Mexico City involve pumping in water from even farther away than is the case today which is neither sustainable nor affordable. The rainwater harvesting (RWH) that Isla Urbana promotes can fill this gap in urban supply and rural access while being both sustainable and affordable.

8.5 SOCIAL ENTREPRENUERSHIP SOLUTION MARKETING APPROACH

After years of testing in our living laboratory, Isla Urbana has developed a RWH system that works with existing infrastructure to efficiently capture and clean rainwater for household uses which is also easy to install and maintain. In addition, we have developed an implementation model to ensure proper use and acceptance of this technology.

The economic opportunity of our RWH systems is in removing the cost of current water delivery systems for families, businesses and governments. In water scarce neighborhoods in Mexico City, families often must spend time and up to 20% of their income to buy water from a truck and municipalities rack up costs subsidizing the trucks. A RWH system that can be paid off in four and a half years is therefore a viable and attractive investment for governments that are required by law to provide their citizens with water service.

We have chosen a hybrid for-profit/non-profit structure in order to reach the most people. Within the non-profit, Isla Urbana has chosen to geographically focus much of our work in water scarce low-income neighborhoods in the south of the city which rely on water trucks despite their above average rainfall. These municipalities wanting to take advantage of the savings from RWH, have sought out Isla Urbana because of the recognition that we have already received for our innovative concept. In rural areas, Isla Urbana seeks out the most marginalized areas which are typically home to indigenous communities. Within the social enterprise, the products have been developed for the specific price levels of low- and middle-income people. We anticipate that in the future, the evidence of success from installed systems together with growing recognition will catalyze city and country wide adoption.

Relationships with RWH customers are extremely important to Isla Urbana as they are the ambassadors of this alternative to standard water provision. Beyond installation, Isla Urbana provides educational activities and resources to promote understanding of our systems and remains in personal contact with households to obtain the feedback necessary for our continual improvement. Isla Urbana partners consistently with universities and institutions to study water quality and investigate impact and with other NGOs when entering new communities. The close relationship we have with suppliers is shown by the fact that one has chosen to donate parts of their profits to the non-profit side of the projects in order to install more systems.

An important determinant of our success is the choice to work with communities or individuals that are the neediest or that approach us. We also require buy-in from beneficiaries through co-financing of the project or at minimum participation through work. This ensures that our systems will be used and will have a definitive positive impact. Additionally, the cultural, social, and educational programs promote the paradigm shift we seek to sustainable water management.

The current water system is very unsustainable, but with centuries of investment and inertia, also very hard to quickly change. Our approach is based on the idea that single rainwater systems provide immediate benefits to individual households, but when replicated in large numbers can have more systemic impacts, increasing resilience and reducing the need to extract or import water from other sources. These systems offer a paradigm-shifting approach to water management, transferring more control from the centralized model to the population itself. Such fundamental shifts would be impossible if they required enormous initial investment and support to work, but the fact that a single rainwater system is in itself highly useful to its users means that progress can be made at any scale, starting with single water-scarce households whose small impacts add up to systemic effects. This allows a potential strategy to build sustainability despite a difficult political and economic context.

8.6 GOALS AND EXPECTED IMPACT

The goal of this initiative is to 'detonate' widespread adoption of RWH in Mexico by installing 100,000 systems at homes, benefitting around 700,000 people. If half of these systems were installed in the city and accounting for leaks in the grid, water extraction would reduce by 3.5 billion liters/year as well as eliminate the energy, emissions and cost required to transport that amount. This would also save municipalities 12 million USD by the end of the three-year period by displacing water trucks. Less quantifiable is the increased resilience of marginal neighborhoods since widespread RWH adoption leaves more water in the grid, lowering the risk of water cutoffs. In rural areas, health and sanitation would be improved by a clean water source. Economic and educational opportunities would increase as the time spent collecting water would be eliminated during the rainy season. Additionally, the training of technicians in designing and installing these systems will provide green job opportunities.

Fundamentally, our goal is to make a serious contribution to water sustainability and equality in our country and city. We believe that effective rainwater harvesting technology and implementation strategies will play an important role in the transition towards sustainability. In the face of the water crisis we are experiencing, our initiative will ensure that the most affected populations in the city have an effective, well adapted and locally controlled means of obtaining water for themselves.

Over the past years, we have been developing and testing rainwater-harvesting technologies in some of the most water stressed areas of the city. These systems have been closely monitored and studied and have proven to be highly effective. Thousands of families are already using them to obtain large percentages of their total water supply. We have been simultaneously developing implementation strategies and collaborating with local governments and universities to study and install them in larger numbers. In order to achieve our larger goals and impacts, we intend to further develop the technologies themselves, reducing their cost and creating more flexible options for people to acquire them. We will continue our work in community and local government outreach to bring the practice to more areas and continue monitoring and evaluation to further demonstrate its effectiveness and potential. Developing financial tools to help low-income families acquire them will also be key.

Currently, local governments have been hiring us to install large numbers of systems in areas with poor water access. This is a great avenue to achieve our goals, but we do not want to depend heavily on government subsidy. Direct sale to users is part of our model, but so far it is more limited to middle- and upper-income families. We want to develop marketing and financing strategies to make rainwater systems fully accessible to the low-income populations most affected by water scarcity.

8.7 IMPLEMENTATION PLAN

Mexico City faces a daunting challenge in achieving water sustainability. The progressive deterioration of the aquifer and surrounding watersheds impose critical questions on how the city will meet its future water needs. Current proposals consist almost entirely on finding and pumping water from farther away, while treating the rainwater that falls on the city as a problem to be solved by expanding the sewage system. Isla Urbana's long-term goal is to develop RWH into a viable, replicable, and proven alternative, able to provide clean water by taking advantage of a wasted resource. By designing and installing systems adapted to the Mexican context, teaching people to do the same, and successfully providing water to areas where the conventional model is failing, we seek to detonate a process of adoption of RWH which will grow as the current system becomes more problematic. As the city fills with RWH systems, the rains that fall upon it will fill its millions of cisterns with clean water, instead of flooding it with combined sewage. As adoption spreads, a wasted resource will become an important source of water, and help the city become more sustainable and less dependent on dwindling, increasingly contested outside sources.

8.8 CONCLUSION

The transition from an intensive unsustainable model of water management to one better integrated with natural water cycles and local availability is not easy and will not likely be sudden in a city as massive and complex as Mexico. Still, the extremely alarming prospects of the city depleting its aquifer and the growing hope for a more rational and sustainable model of water management are creating both the need and opportunity to explore real alternatives.

Mexico City needs to develop new sources of water, as well as to find ways of using it in more rational ways. Even by optimistic projections where the city implements all recommendations to reduce its profligacy, a quarter of water needs in 2030 will have to come from undefined new sources. The geographic location of the Valley of Mexico, 2200 m above sea level and far from any major rivers or lakes, makes this a daunting challenge. And yet the city has an abundant source of water naturally transported directly to it though the yearly torrential rains that fall over the valley.

Currently this rainwater is mostly drained out of the valley through the sewage system, becoming intensely polluted and causing widespread flooding. With proposals for increasing supply to the city consisting largely in grandiose, hugely expensive plans to pump water from far greater distances, the possibility of learning how to use the rainwater within the valley must be explored in earnest. This is the mission of the Isla Urbana project which, since 2009 has been developing technologies and implementation methods to adapt Mexico City's houses to harvest and use the water that falls on their roofs.

By demonstrating that houses in the city can viably be made water independent for much of the year though the adoption of simple harvesting technologies, Isla Urbana seeks to develop a truly alternative, sustainable form of water management that can add resilience to the city and reduce the need to import or extract water on such massive scales. It has been working to develop a water-stressed section of the city into a first example of widespread adoption of household rainwater use where all the potential benefits can be shown and questions on how this can be achieved can be answered.

In collaboration with the Secretary of the Environment of Mexico City, Isla Urbana is currently working to bring this pilot to full development. It has a goal of installing 100,000 RWH systems in the area over the next four years, enough to supply water for much of the most water stressed population, while carrying out research and evaluations of their impact. If successful, this effort will demonstrate the potential of using RWH as a means of adding resilience and sustainability to the city's water management strategy and provide the practical knowledge on how to bring it to reality.

REFERENCES

Banco Mundial, CONAGUA (2013). Agua Urbana en el Valle de México: ¿un camino verde para mañana? World Bank, Washington, D.C.

Isla Urbana (2018). "Pluviateca", http://islaurbana.mx/pluvioteca/

Tortajada, C. (2006). "Water Management in Mexico City Metropolitan Area", International Journal of Water Resources Development, Volume 22, Issue 2, pages 353–376. http://10.1080/07900620600671367

Chapter 9

Promotion of rainwater harvesting as a business

A. Kikundwa Birungi

Liaison Officer, Uganda Rainwater Association, Kampala, Uganda

Keywords: best practices, climate change, economic growth, environment conservation, health, integrated water resources management, production, rain for sales, rainwater harvesting, safe water

9.1 INTRODUCTION

Without access to safe water, people, especially those in rural areas, are trapped in a cycle of poverty and disease. Women walk long distances to collect water, which reduces their involvement in productive work for their households. On some days, children forego school to collect water for home use. Water and sanitation-related diseases such as diarrhea and typhoid are common.

Rainwater harvesting is a viable solution to the access to water situation; the water is within the homestead, and if collected and stored properly it is free from major contaminants. It brings tremendous benefits in terms of reduction in water-borne diseases, improved school retention and increased productive time (for women) due to water-collection time saved. As such it contributes to the fulfillment of the SDGs related to health, poverty reduction, gender and education, as well as meeting the country's National Development Plan that identifies provision of adequate water supply and improved sanitation as one of the key strategies for promoting economic growth and reducing poverty.

Because of high poverty levels, households are unable to raise money to procure rainwater-harvesting systems (let alone other amenities) on their own. As a result,

 International Rainwater Catchment Systems Experiences: Towards Water Security
Editor(s): José Arturo Gleason Espíndola, César Augusto Casiano Flores, Raul Pacheco-Vega and Margarita Rosa Pacheco Montes
doi: 10.2166/9781789060584_0119

several households are involved in community groups, dealing in saving and loaning for various purposes, but often not for investment or water provisioning. Most of these groups are registered and recognized (as community-based organizations, CBOs) by local governments in their areas of operation. These groups are an excellent entry point to facilitate self-supply of rainwater harvesting facilities for economic development, and this is illustrated further in our case in this chapter.

Rainwater harvesting is a simple and low cost water supply technique that involves the capturing and storing of rainwater from roof and ground catchments for domestic, agricultural, industrial and environmental purposes. When surface runoff is collected in reservoirs, it can be used for the management of floods and droughts (environment conservation). Surface runoff can also be used for recharging groundwater, which will positively impact on springs and shallow wells. Rainwater harvesting yields numerous social and economic benefits, and contributes to poverty alleviation and sustainable development.

In Uganda, rainwater harvesting is increasingly becoming important both for agriculture and domestic water use. This is due to increasing drought spells, erratic rainy seasons which cause poor agricultural yields resulting in negative impact on the country's economy. Some rainwater harvesting technologies are being promoted both by the Government and NGOs in several parts of the country. These range from water storage techniques at micro and macro level e.g. small scale dams, underground and above ground tanks to in-situ technologies e. g. soil and water conservation technologies. Farmers, Government, Donors and NGOs have invested huge resources (land, labor, and capital) in order to harvest water using different techniques.

Promotion of Rainwater harvesting for sustainable livelihoods is the main focus of the Uganda Rainwater Association (URWA). URWA contributes to integrated water resources management (IWRM) through promotion of efficient management of rainwater. Rainwater harvesting as a business has been a neglected component of water resource management yet it has a direct impact on people's livelihoods. This recognition presents new challenges which need to be addressed and opportunities to be leveraged. Against this background, URWA embarked on empowering entrepreneurs with skills and knowledge to be able to make more informed decisions and choices about managing rainwater for their wellbeing. As the impact of climate change manifests itself, we tighten our belts to pay serious attention beyond the conventional sources of water and look towards integrated rainwater harvesting techniques to boost our business ventures.

9.2 RAINWATER HARVESTING AS A BUSINESS

Water is precious. We cannot shy away from the fact that pressure and the demand on the water resource grow day by day. Therefore, there is a need to utilize the best practices of rainwater harvesting to make sure that there is no water shortage in the future. Water is society's most basic need; without it nothing can live. It is vital for health, production, and economic growth. Access to clean, safe and reliable water supply is a universal human right. Although Uganda is endowed with water in terms of rainfall and water bodies, this water is not accessible to all in the right amounts and quality at the right time. We have been warned several times by climatologists that there are difficult times ahead because of climate change. As the climate warms, rainfall will become more erratic with severe floods, and droughts will be more frequent, longer and more severe with catastrophic consequences, thus the need to seriously promote rainwater harvesting and venture into business for economic growth.

9.2.1 A Case of the Rainwater4Sale project in Lwengo District, Uganda

Rainfall variability has emerged as one of the most challenging among climate change events for communities in general. In this regard, the lifestyles of different social groups can be negatively affected from time to time by excessive rainfall and extended drought, caused by rainfall variability. The small farmers and the households in Lwengo District are extremely vulnerable to these changes as the entire wellbeing of the farming community is completely dependent on the capacity of these communities to access the right amounts of water at all times to ensure yields from the crops and livestock. In Lwengo, during the dry period, farmers lack water to meet the needs of the farm. An imperative for rain fed farming therefore, is that a technology or innovation to store water for later use when needed. The percentage of households using rainwater harvesting in rural areas of Uganda is low. Moreover, with increasing populations and high unemployment there is more need for water for domestic use, production, and other purposes to enhance economic development. Rainwater harvesting has the potential to provide water for these purposes, and improve food production for communities who have a high dependence on agriculture.

Against this background, a Rain4Sale pilot project was carried out in two sub-counties of Lwengo District. In Lwengo, and Uganda at large, there is a persisting problem of rainfall being "too much or too little" on a seasonal basis. With climate change, the dry season becomes drier and the wet season wetter. This results in uncontrolled surface runoff, soil erosion, and flooding of low lying areas. All these misfortunes impact negatively on the economic standards of community members. There is no piped water in these sub-counties, and these communities were willing to adopt rainwater harvesting to meet their needs.

The aim of this project was to implement water harvesting and conservation techniques that would assist the communities in this catchment area to improve their livelihoods by enhancing their income through sales of stored rainwater for different purposes.

In June 2016, financial support was acquired from RAIN to give to entrepreneurs to build 50 m^3 rainwater harvesting systems in Lwengo District. The rainwater harvesting business model was introduced to four entrepreneurs. They received loans for construction of rainwater harvesting (RWH) installations and are currently paying back in installments. Some reimburse quarterly, while others reimburse monthly.

Rainwater harvesting business model is a sustainable way of initiating a profitable venture. The entrepreneurs collect and store the rainwater into 50 m^3 ferrocement tank (Figure 9.1). They have adopted the best practices of utilizing rainwater for domestic use, agricultural use, and environment conservation. This water is also sold to communities who use it for drinking. The entrepreneurs use the water for livestock (Figure 9.2) and irrigation, and are earning income from the products. In addition to the above, this technology provides social development, economic wellbeing and environmental sustainability. The rain for sales project provided employment to some masons in the community. They were trained and mobilized to implement this technology. Communities are encouraged to mobilize construction materials so as to build rainwater harvesting systems and provide water for domestic and economic purposes. In this project, women confess that the rainwater harvesting technology facilitates them by providing water which is otherwise brought from distanced water sources and reduces their physical hardship and mental stress as well as time required to haul water from other water sources.

The saved time is thus used for other productive purposes such as domestic work, agriculture and livestock activities (Figures 9.3, 9.4 & 9.5), and child care. Installing these rainwater harvesting systems reduced the water supply costs and also provides significant savings.

URWA's focus was initially entirely on rain water harvesting, collection and re-use, but today it goes well beyond the selling of rain water. We encourage entrepreneurs to re-use the rain water for production, and sanitation improvement.

Figure 9.1 50 m^3 ferrocement tanks. (*Source*: Kikundwa Anne).

Existing surface water and groundwater resources are being depleted. Already, there are places in Uganda that are experiencing shortages because demands are greater than available supplies. Rainwater harvesting therefore provides us with an opportunity to conserve and extend our existing resources. It is important to note therefore, that rainwater harvesting is not intended to replace other conventional sources of water. Harvesting rainwater simply provides an additional measure of security to communities during times of water scarcity.

We continuously interact with entrepreneurs to ensure best practices of operation and maintenance, client satisfaction issues, and aspects of the safe water chain. Quarterly monitoring is ongoing. The entrepreneurs report any issues that come up regarding the business.

Figure 9.2 Piggery. (*Source*: Kikundwa Anne).

Figure 9.3 Brick laying. (*Source*: Kikundwa Anne).

Figure 9.4 Kitchen gardening. (*Source*: Kikundwa Anne).

Figure 9.5 Passion fruit growing project. (*Source*: Kikundwa Anne).

9.2.2 Lessons learned

Different people may have the same problems, but they need to be addressed differently because what works for one person may not necessarily work for another. This has been witnessed amongst the different entrepreneurs who had a common goal of making money out of rainwater sales, but used different approaches in the utilization of rainwater as a business.

Collaboration of all stakeholders is crucial because different actors have different important roles that will contribute to the success of projects. Successful projects are not a 'one man' business, but a result of combined efforts from all stakeholders.

For inspiring change, partnerships should be emphasized. This helps the concerned parties to make informed decisions and perform better.

Not all plans flow as well as predicted. There are ups and downs in project implementation and these require going back to the drawing board to develop appropriate strategies.

9.2.3 Challenges and how they were addressed

We have had issues when it comes to loans reimbursement. Not all the entrepreneurs were repaying on time, or sometimes a repayment period would find them with another period's debt. We sat with the entrepreneurs and to lay out strategies on how to address this issue.

Record keeping was a challenge. This was addressed by introducing Journal Vouchers where entries of how many jerrycans sold, and the amount are recorded on a daily basis.

During the dry season, water drains away quickly and the customers get back to their conventional sources of water. Sometimes they queue at his home thinking that he is refusing to sell the water to them.

During the rainy season, the costs per jerry can are low, and the customers are fewer. This reduces the sales, but the entrepreneurs were encouraged to store the water and maximize sales during the dry season.

9.3 CONCLUSION

Rainwater harvesting is a simple, inexpensive technology that promotes sustainable water management. This technology should be adopted for economic development. Lessons from the Rain4Sale project testify to this. When projects are implemented, stakeholder involvement in the project does not end at the project phase-out stage but must continue through and through for sustainability and continuity. We have heard of projects that die due to lack of monitoring and follow ups. In this rain for sales project, lessons drawn need to be used to inform other projects that may come up, and also address issues that may come up and hinder the smooth running of the rain-for-sale business. There is so much interest of late in rainwater harvesting. Rainwater harvesting is enjoying a revival in popularity.

We continue to spread the self- supply gospel to embrace promotion of rainwater harvesting as a business.

WHO WE ARE

The Uganda Rain Water Association (URWA) is a membership organization founded in December 1997; it was registered in 1999 as a Company Limited by Guarantee followed by a 2006 registration as a local NGO, to raise rainwater management and utilization among all development sectors in Uganda. URWA supports communities to improve their socio-economic situation through mobilization, information, skills and experience sharing and collaboration between its members. The major focus of community support is integrated rainwater harvesting and management for improved livelihoods. They are facilitated to promote rainwater harvesting for domestic use, environment conservation and production.

OUR MANDATE

We support communities to improve their socio-economic situation through rainwater harvesting for domestic water supply, agricultural production, and environment use. We do this through mobilization, sharing information, advocating, building capacity and researching on innovative technologies and approaches of harvesting and managing rainwater.

Chapter 10

Experience in sustainable management of rainwater for multiple purposes: Case in ten villages, gossas district, Senegal

S. Souleymane

Caritas Kaolack, Senegal
sene.sgs@gmail.com

Keywords: bacteriological, chemical, freshwater, harvesting, groundwater, management, rainwater, Senegal, sustainable

10.1 CONTEXT

Senegal aims to become an emerging market by 2035. The Senegalese Emerging Market Plan (*Plan Sénégal Emergent; PSE*) has guided this ambition since its inception in 2014. The PSE is the main socio-economic reference document for the country's medium- and long-term development. The plan's 'Human Capital: Social Security and Sustainable Development' section identifies access to drinking water as being integral to achieving Senegal's socio-political development goals.

Since 2015, efforts to provide these basic social services have accelerated, in conjunction with the programme outlined in the Plan for Urgent Community Development (*Plan d'urgence pour le development communautaire;* PUDC), in which water infrastructure plays a key role.

Programmes including Pepam Aqua, a Senegalese-Belgian collaboration, have enabled Senegal to explore different avenues for creating potable water supplies, from brackish water extracted from boreholes, and by harvesting rainwater. The rainwater purification project at Walalane results from Pepam Aqua's collaboration with Caritas Kaolack.

 International Rainwater Catchment Systems Experiences: Towards Water Security
Editor(s): José Arturo Gleason Espíndola, César Augusto Casiano Flores, Raul Pacheco-Vega and Margarita Rosa Pacheco Montes
doi: 10.2166/9781789060584_0127

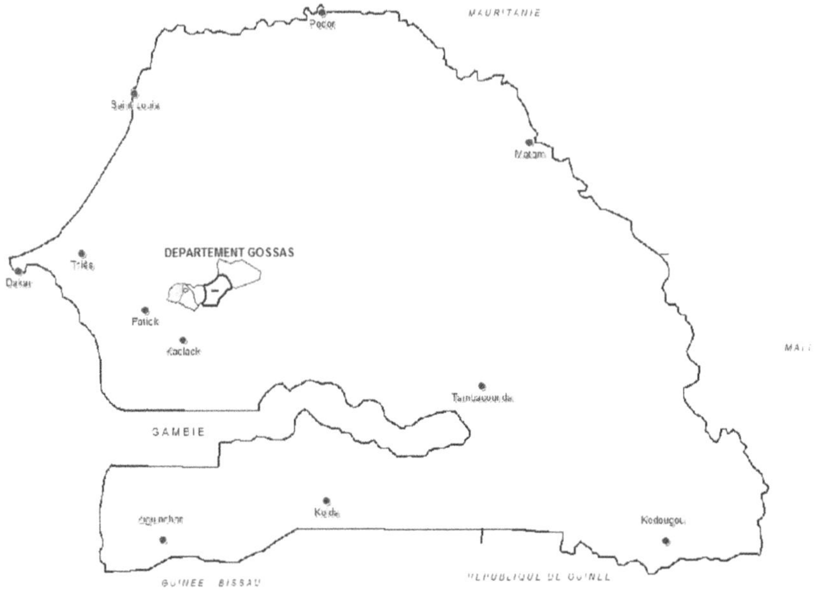

Figure 10.1 Project zone situation in Gossas. (S.Sene, 2018).

Since 2006, Caritas Kaolak have instigated multiple projects, with the support of the Dutch NGO, Rain Foundation, to harvest rainwater as a drinking water supply for communities living in regions where surface and groundwater freshwater supplies are scarce. In such locations, it is sometimes necessary to canoe for tens of kilometres to access a drinking water supply, which might be prohibitively priced. For inhabitants of the Saloum Islands, acquiring a 20 l barrel of drinking water costs 0.3 euros (15 euros/m^3).

Setting up rain-harvested water supplies for school children in Senegal's Gossas district (Figure 10.1), with project funding from Horizont3000 (Meerman et al., 2015), aligned with Senegal's national development policy: to make drinking water supplies available to its populace, irrespective of where they live. Children are the future of Senegal, and providing this vulnerable sector of the population with water through such projects is of vital importance.

10.2 HYDRO-GEOGRAPHY AND HYDRO-CHEMISTRY IN GOSSAS

Gossas, located in the extreme north of the Fatick region, experiences a semi-arid, Sahel-Soudanian type climate, with annual precipitation levels rarely exceeding 500 mm. Vegetation in the area is characterised by spiny shrubs and members of the *combretaceae* family.

Most water resources in the district derive from deep groundwater reserves. Wells supplying freshwater are dug to minimum depths of 70 m below ground. However, a Maastrichtian-age aquifer, located between 200–400 m below ground, contains the best water reserves. Boreholes used to access this drinking water resource typically supply 50 m^3/hour.

A key priority of the Senegalese government is to supply drinking water to all its inhabitants. Given that 45% of Gossas district's rural population lacks access to drinking water, developing a range of water supply options for this demographic clearly contributes to meeting their water needs (CONGAD, 2014). Provision of sanitation facilities in rural regions such as this one is, similarly, of great importance: on average, only 49.1% of the population could access such amenities in 2013. In the rural district of Mbar, access to sanitation was only available to 7% of the local population (PEPAM: *Projet Eau Potable et Assainissement pour le Millénaire*).

Survey work was conducted before boreholes were drilled by PEPAM – Bassin Arachider (Project Drinking Water and Sanitation for the Millennium – Groundnut Basin). Results from this survey demonstrate that out of 53 boreholes surveyed in the districts of Diourbel, Fatick, Kaolack and Kaffrine, 32 supplied water whose quality did not meet with the standards defined by the World Health Organisation (WHO). These waters had excessively high concentrations of fluoride, salts and, in some locations, iron (Initial Field Survey; PEPAM-Aqua (2011a)). For instance, fluoride levels of 4–5 mg/l were sometimes measured in the field; these levels far exceed the 1.5 mg/l stipulated by the WHO. Similarly, salt levels in excess of 1.5 g/l were measured; WHO recommends that salt levels be less than 250 mg/l in drinking water (National Hydraulic's Direction, 2015).

Water analyses conducted by Caritas, deriving from three boreholes drilled in the Gossas district, where the PEPAM–BA project is located, find comparable levels of chloride and fluoride.

Water samples systematically contain double or triple the concentrations recommended by the WHO for drinking water, human consumption of such water sources is harmful (Table 10.1). Public health surveys in the Patar Lia and Ouadiour local communities within the Gossas district report cases of arterial hypertension, stomach pains, diarrhoea, and skeletal fluorosis in pregnant women, children and old people that regularly consume water from boreholes.

Table 10.1 Physico-chemical analyses of some boreholes in gossas.

	Soumbel Keur Latyr Borehole	Gossas Village Borehole	Sakhmack Borehole	WHO Recommended Concentrations
Chloride (mg/l)	475.2	550	404.2	250
Fluoride (mg/l)	4.5	3	5	1.5

Given the high levels of mineral salts in this water source, local communities increasingly reject it, choosing to walk long distances to obtain safe drinking water, or using water stored in traditional wells that are restocked with rainwater, which is often of poor quality.

Groundwater supplies might appear limited, over-exploited or difficult to access in Senegal. However, in the rainy season, millions of cubic metres of rain are recorded as falling. Much of this precipitation is lost as runoff that enters the sea. Harvesting rainwater is increasingly relevant in Senegal, given the decreasing groundwater resources and groundwater quality in many areas, and the cost of drilling boreholes that yield water that is often not of a potable standard.

Furthermore, water access in school contexts is often particularly problematic, given the difficulty of connecting these institutions to a supply that can be drunk, used to maintain hygiene and sanitation facilities, and thirdly employed to improve the quality of school learning environments by irrigating vegetation that can provide shade and beauty. While precipitation is not abundant in the Gossas region, it is of sufficient volume to satisfy the drinking water needs of local communities, and to improve the learning environments of school children.

10.3 METHODOLOGY

This project was developed in collaboration with farming organisations in the Gossas District, including the *Association régionale des Agriculteurs de Fatick* (ARAF). Fatick's regional association of farmers is well represented in the Gossas District, and local members have a good understanding of the needs of each village, particularly with regard to their potable water supply requirements.

Collaborations were developed with the local mayors in villages in the Gossas district, to guide their decision making with regard to the following issues: the existence of primary schools lacking a supply of freshwater; the acceptance of rainwater harvesting as an alternative potable water source in this context; a willingness to experiment with school gardens on campus; an openness to local authorities offering in-kind support of projects by supplying local materials for building projects and local labour; and teaching local masons how to construct rainwater harvesting systems.

In all of the target sites, field visits and meetings were conducted to ensure the project's sociological and technical feasibility. The schools involved in the project were visited to gather data about the school populations and to study how rainwater could be collected from each school's roof; to consider what size of the rainwater harvesting tank would be needed, the quality of the roofing materials, and how to store drinking water reserves, while distancing them from the toilet facilities.

In each village, a kick-off workshop was organised at the beginning of the project, and before the village councils developed a strategy for implementing it. These workshops helped participants identify collaborators' roles and responsibilities, how they would participate in the project, and project deadlines.

The volume of water needed to supply the drinking water requirements of each school for a year, assuming 1.5 litres per person, per day, was calculated. The number of school children and staff in the project's 10 villages totalled 1368. The combined population of the 10 villages involved in the project was 5365. Accordingly, it was decided that 100 m^3 of rainwater (stored in two 50 m^3 systems) would meet the water requirements of each school, and give the communities associated with these schools punctual access to this water resource as well, to test its quality with a view to scaling up rainwater harvesting in the villages.

Since one of the project aims was to create green zones in these village schools, by planting trees for shade and growing vegetables for four months of the year, an additional 60 m^3 cistern was constructed specifically for irrigation purposes. It was estimated that 24 litres per day would be required to irrigate eight raised beds that were each 1 m^2; 200 litres per day were also allocated for the other watering purposes. To calculate the required surface area of the rainwater cisterns, the following formula was used:

$$\text{Cistern surface area} = \text{total volume of estimated rainwater}\,(\text{m}^3)/\text{precipitation}\,(\text{m})$$

The area calculated was increased by 10%, to compensate for possible water losses.

The reservoirs constructed for this project (Figure 10.2) were partially buried, with 2 m of the structure located below-ground and 0.5 m protruding above ground. These tanks collected water from the roofs, using corrugated iron or aluminium gutters, at the end of which a funnel covered with a metal mesh was attached. This semi-permeable barrier performed a initial filtering of the rainwaters before they ran, via a PVC pipe, into the rainwater reservoir or cistern.

Figure 10.2 Rainwater harvesting system in Dekhaye School. (S.Sene, 2018).

Rainwater is collected once the season's first rainfall has occurred, which flushes dust from the roofs and piping. From the second rainfall event, (equivalent to or greater than 20 mm), all rain falling onsite is collected in a partially submerged tank, for use as a potable water resource. Once a management committee is put in place for the rainwater harvesting system, its role includes ensuring that the roofs that collect the rainwater are kept clean of debris.

10.4 RESULTS

30 rain cisterns were constructed. 20 cisterns were 50 m^3, nine were 60 m^3 and one was 40 m^3, enabling 1000 m^3 of rainwater to be collected for drinking water purposes, and 580 m^3 to be collected for gardening purposes, in total.

In one village involved in the project (Walalane), Caritas had built a large (1400 m^3) rainwater harvesting tank in the 1980s, enabling the local village women to continue market-gardening during the dry season for 3–4 months. This water tank had begun to leak. It was re-sealed and returned to its original use, enabling market gardening in Walalane to be restarted.

In each of the ten schools involved in the project, eight raised beds were built and installed. Responding to the wishes of each school and its neighbouring village regarding what vegetables they would like to farm, on-site nurseries were established, using honeycomb-shaped planters. To enhance the longevity of this gardening strand of the project, school children and staff responsible for maintaining the schools' outdoor environments were actively involved.

Two types of agricultural substrate were adopted for these raised beds: a mix of well-washed gravel, peanut shells, and rice grain husks; and hydroponics, using 80 litres of water containing macro- and micro-nutrients.

In total, eighty 1.2 m^2 raised beds were planted. 39 of these beds used a solid substrate and 41 used specially constructed hydroponic systems.

Ultraviolet light was used to purify the water in each school, thereby removing live bacteria. A small, solar-powered pump enabled water to be pumped from the reservoirs through this system.

Laboratory analyses were undertaken to compare water samples obtained directly from the rainwater harvesting reservoirs and samples of water that had been exposed to UV light.

10.5 DISCUSSION

The results of the bacteriological analyses (Table 10.2) show that water obtained from the different rainwater harvesting systems does not present any major risks for consumption (with the exception of the Tchingué site), if adequate treatment with UV light is adopted. Ultra-violet light is effective in eliminating coliform and *enterococcus* bacteria, which provoke gastroenteritis, if basic hygiene measures are undertaken in parallel with this water treatment measure. Regarding

Table 10.2 Bacteriological analyses from rainwater harvesting tanks in different villages (Sene, 2018).

Parameters Measured	**Water in the Cistern**	**Water Exiting the UV Lamp**	**Treated Water Exiting the Reservoir**	**Drinking Water Norms (WHO)**	**Water in the Cistern**	**Water Exiting the UV Lamp**	**Treated Water Exiting the Reservoir**	**Drinking Water Norms (WHO)**
		Mbar Bawane				**Boustane Diaw**		
Germs: Total (UFC)	202	22	648	20/1 ml	162	50	58	20/1 ml
Coliform bacteria (UFC)	310	0	0	0/100 ml	0	0	0	0/100 ml
Enterococcus bacteria (UFC)	460	0	0	0/100 ml	920	0	0	0/100 ml
		Tchingué				**Dékhaye**		
Germs: Total (UFC)	520	>**880***	680	20/1 ml	90	34	976	20/1 ml
Coliform bacteria (UFC)	60	**58**	**50**	0/100 ml	0	0	0	0/100 ml
Enterococcus bacteria (UFC)	1130	**312**	**80**	0/100 ml	140	0	0	0/100 ml
		Bill Mbacké				**Loumbel Kelly**		
Germs: Total (UFC)	815	–	174	20/1 ml	184	–	120	20/1 ml
Coliform bacteria (UFC)	0	–	0	0/100 ml	10	–	0	0/100 ml
Enterococcus bacteria (UFC)	150	–	0	0/100 ml	10	–	0	0/100 ml

the Tchingué site, this location had a water cistern that was always open for drinking water consumption. The water it stored was unpotable, even after it had been treated with UV light, due to the continued presence of *enterococcus* bacteria. Consequently, this rainwater harvesting tank was only used, subsequently, for irrigation purposes.

Even if water becomes potable after it has been treated with UV radiation, appropriate hygiene protocols need to be put in place, to prevent the water from being re-contaminated. The laboratory analyses show that when dust falls on the water reservoir barrel, if they are not well sealed around the entry point of the rainwater pipe, further contamination of the water source can occur. Accordingly, the vigilance of those managing and using harvested rainwater resources is paramount to maintaining water quality. A user's guide was made to teach project participants what precautionary measures to take when consuming water that might be contaminated. This included using chlorine to treat the water, or adding bleach at 8°C.

Using water from the cisterns, the schools were able to irrigate their mini-gardens and harvest produce including lettuce, tomatoes, turnips, aubergines, carrots and mint (Figures 10.3 & 10.4). This herb is much used locally in tea, and in drinks such as *bissap* (*Hibiscus sabdariffa L.*) or red Guinean sorrel.

Additionally, as a result of the variety of vegetables that were harvested (salads, carrots, turnips, okra, sweet and bitter aubergines, and tomatoes) the project beneficiaries were able to improve the nutritional value and diversity of their meals during at least three months. The beneficiaries' households had never previously been able to grow or consume these vegetables out of season, and, with regard to carrots and turnips in particular, certainly never in the volumes produced by the project.

Figure 10.3 Development state of tomato and eggplants. (S.Sene, 2018).

Figure 10.4 Eggplant harvest after 2.5 months. (S.Sene, 2018).

Growing crops such as tomatoes considerably increased the revenue of project beneficiaries. As an example, in the village of Walalane, where market gardening was conducted in fields, and irrigated using the above-mentioned restored water tank, 604 kg of tomatoes were harvested from a 100 m^2 parcel of land and sold for 22,500 Fcfa (343 euros). This money was used by the five, female project beneficiaries, to improve the quality of the food they bought for their families, by buying seasonings and spices and additional food stuffs.

Rainwater harvested at Walalane was sold, generating a revenue of 1,125,000 Fcfa (1715 euros) over a period of nine months, in a year when rainfall was abundant. It is estimated that wages for the manager of the tank cost 225,000 Fcfa (343 euros), representing a net gain of 900,000 Fcfa (1371 euros). Given that constructing a rainwater harvesting reservoir and water purifying system of the type used in this project costs 5,000,000 Fcfa (7622 euros), the money gained through water sales would enable a new rain cistern to be built every six years.

The initial purpose of the Walalane rainwater harvesting initiative was to meet the water needs of this village's school and community, in part if not completely. However, water was sold to people traveling from the towns of Gossas and Diourbel, and also neighbouring villages on a punctual basis, when water canisters were available. These consumers considered that this rainwater was of a better quality than the groundwater available to them in their towns and villages.

Improving school environments was a final, important element of this project. Planting trees to provide shade is valuable in a context where the dry, hot

Harmattan wind, which blows between February and May, makes it uncomfortable for students to spend time outside. Students are invariably obliged to remain in class during their breaks, even in the intense heat of the afternoon, because it is even more uncomfortably hot outside. Harvesting rainwater allows plants to be watered during the rainy season and through the whole of the dry season. In less than two years, trees cultivated in this manner can grow from 50 cm high saplings to 2 m high tree species. This vegetation transforms the environment of school playgrounds, giving school children shady outdoor spaces in which to play.

10.6 DIFFICULTIES AND LIMITS OF THE PROJECT

In the project's first year (2017), implementation of the rainwater harvesting infrastructure coincided with a period of very little rainfall. The little rain that did fall between July and August was not able to be collected, as the cisterns were not yet completed. Consequently, the volume of rainwater that was harvested did not meet the local demand for water.

Additionally, some gutters did not function well, either as they had been damaged due to winds and rain, or because the management and maintenance committees in charge of the rainwater harvesting infrastructure had not sufficiently taken care of them.

It is normal for households with a reasonable revenue to pay 150 F for 20 litres of water. However, for other households, buying 40–60 litres per day to meet the water needs of a large family proves to be a costly expenditure, constituting a large portion of the household budget. By identifying which groups were most in need of free access to harvested rainwater, in this context, specifically pregnant women, children and old people, the harvested rainwater was used most effectively within local communities. This is not to deny the relationship between health and water, but rather to make sure that at-risk groups had their minimum requirements for water met.

Water cisterns and tanks were built in locations exposed to the easterly Harmattan wind were observed to become rapidly re-contaminated, even as the water was purified, due to dust and pollution becoming deposited in holes and crevices on the rainwater harvesting cisterns and associated structures.

10.7 LESSONS LEARNED

Rainwater is a useful alternative water source that can be used to provide potable water and water for irrigation in arid regions with low precipitation. Thus, while rainwater does not conform to the norms of water collection, harvesting and consumption, it is a good quality drinking water resource. Harvesting rainwater for these purposes can resolve problems of water access, experienced by populations trying to access more traditional drinking water resources. However, this approach requires that the appropriate infrastructure and equipment is put in place, and that hygiene measures are scrupulously respected.

The quality of water coming from rainwater harvesting systems was appreciated by all project beneficiaries in our endeavour, as it improved their health, particularly with regard to illnesses such as arterial hypertension and bone pain, which pregnant women and the geriatric population are particularly prone to suffering, if they consume brackish waters or waters highly concentrated in fluoride. That said, it is paramount that the water quality is regularly assessed in rainwater harvesting systems; testing should be conducted on rainwater as it is harvested, stored and distributed.

Poor drinking water quality can often be traced back to a water's source. But it is also important to assess the quality of water as it is stored and distributed, as water can be re-contaminated in both these phases. Those using rainwater as a drinking resource should always be careful in storing rainwater, and they should take appropriate measures in treating water and in handling it hygienically.

The collection and use of rainwater in agricultural settings can favour the diversification of farming practices in regions where climate change poses challenges to production. This is a wise use of a resource that is often lost, as it runs off into the sea, without having been utilised.

Harvesting rainwater in school contexts can significantly improve the quality of life of students. This resource can be used to cultivate trees and clean toilets. Consequently, the school environment evolves rapidly, improving the learning conditions of school children and their teachers, thereby allowing young scholars to blossom.

Creating micro-gardens using harvested rainwater can also extend the range of fresh vegetables available to a community. Many of the vegetables grown as part of this project were never previously available to beneficiaries. Additionally, the creation of these micro-gardens had a non-negligible effect on the quality of practical teaching methodologies. Previously, home economics classes had been predominantly theoretical; with the implementation of the micro-gardens, practical classes are now also offered by teachers.

Micro gardens provide the best solution for cultivating vegetables in school grounds, where water resources are often limited. As they consume little water, it is possible to irrigate them throughout the dry season. Micro garden cultivation gives schools a way of earning money through selling produce. However, it is important that the practical requirements associated with tending a micro garden are met on a daily basis. Their management requires the commitment of school children and their parents, who provide invaluable help in maintaining school gardens, particularly during school holidays.

For rainwater harvesting systems to function optimally, it is necessary for them to be regularly maintained. Therefore, a management committee needs to be established, who can pay particular attention to preparing the systems before the first rains of the season. This is the only means of ensuring that rainwater cisterns are filled effectively, and with water that is of potable quality.

This project has demonstrated the pertinence of rainwater harvesting in Senegal, and in the Sahel region more generally. Rainwater harvesting has accordingly been adopted by the FAO, who, in 2018, inspired by this project, launched the initiative '*un million de citernes pour le Sahel*' (a million of cisterns for the Sahel). This project was supported by Caritas, who made personnel and resources available during the pilot phase of this initiative in Senegal and Niger.

10.8 CONCLUSIONS

This project is a new initiative in this region; accordingly, its results could not be anticipated at the outset. Drinking water is not a common activity in this region, not is the consumption of fresh vegetables in the dry season. Accordingly, our project now presents us with the challenge of presenting a new image for rainwater harvesting. This resource is no longer only pertinent to irrigating large areas of winter crop. Today, harvested rainwater can be used to meet other needs, thereby modifying the daily activities of project beneficiaries.

After two years of project implementation, the local population of the Gossas district understands the value of rainwater harvesting for multiple uses. Seven to nine months after the rainy season ends, the project's cisterns still contain rainwater, which can be employed in a diverse manner, and often in applications that are innovative.

What are our governments waiting for, in order to integrate rainwater harvesting into rural and urban hydrological development projects? This is a legitimate question in a context where we know that water resources costing billions of euros are presently being developed to improve millions of people's access to freshwater resources; for instance, through desalinisation projects. We know, however, that this type of solution will privatise freshwater resources, and make them increasingly inaccessible for those with low incomes. In contrast, rainwater, manna from heaven, could ensure the democratisation of access to drinking water resources. Rainwater cisterns are easily reproducible, particularly for low-income communities, whose low revenues often preclude their being able to prioritise the purchase of this vital resource. They prioritise the purchase of basic food stuffs, and tend to utilise water resources that are of poorer quality than rainwater; the adverse consequences of consuming such waters will indirectly increase the cost of public health care for this demographic.

Projects such as this one merit more governmental and private sector investment, as well as the financial support of international bodies. They assist populations that might consider themselves as being overlooked by the political system.

10.9 BENEFICIARIES' TESTIMONIALS

> 'When I was still drinking water from the borehole, during my pregnancies, I had troubles with my legs. My feet swelled up to the extent that I had trouble walking,

and I had to go to the toilet a lot to urinate. The nurse told me these symptoms were related to the poor quality of my drinking water, and gave me a tablet to dissolve in the water coming from the borehole before drinking it. Since the rainwater harvesting system has been installed in my village, this is the water that I drink. I have not suffered from the same symptoms during my pregnancies once I made this change. My last child, who I'm carrying on my back, was an example of an easier pregnancy. This water has become really important for us'.

Ndiague Faye, mother, Walalane.

'I live with my mother, who is old. When she drank water from the borehole, she often suffered from hypertension and rheumatism. I took her to the hospital in Gossas, located 7 km from our home, for her to be cared for there, and this cost me lots of time and money. Since our rainwater harvesting system has been installed, and she has begun to drink this water, she is in better health. This is why, even if we have run out of harvested rainwater at home, she doesn't attempt to drink water from the borehole, and rather prefers to go and ask for rainwater from one of our neighbours'.

Bathie Faye, son, Walalane.

'We never thought we'd have fresh vegetables in the dry season, certainly not on our tables, and in such great quantities; it's incredible! Today, in our school canteen, we have different vegetables, which improve the quality of our meals'.

Teaching staff member, Tchingué

ACKNOWLEDGEMENTS

I would like to thank very much Rachel Nisbet, from IRHA H2O Geneva, for all her help, translating and arranging my document.

Thanks also to Margarita Pacheco, for the coordination of the work and always her recall, for the deadline.

And to all the members of rainwater harvesting community, with the headmaster, Dr. José Arturo Gleason Espíndola, President of Mexican Rainwater Catchment Systems Association, who makes this work possible, and allows me to share this small experience.

Last but not least, thanks so much to HORIZONT3000 and its country representative, Ismael Ndao, who provided funds for the implementation of the pilot and scale up of the rainwater harvesting project in Gossas District. I can't forget my colleagues of Caritas, who work hard to make this project and its results possible.

REFERENCES

Caritas Sénégal (2009). Impact of fluoride ingested via drinking water on the health of populations. Case of five rural communes, May 2009. Multidisciplinary Research Team, Synthetic Document, Caritas, Senegal, 6p.

CONGAD. (2014). Document de plaidoyer des ONG du Sénégal. UNDEF, Senegal.

Meerman R., Sène S. and Nitzsche M. (2015). Towards an Integrated Rain Water Harvesting Approach in Senegal, Concept Note Project, October 2015, 13p. Horizont3000, internal document National Hydraulic's Plan (2015) Note on Senegal's water resources, October 2015. National Hydraulics, Senegal.

National Hydraulic's Plan. (2015). Note on Senegal's water resources, October 2015. National Hydraulics, Senegal.

PEPAM-Aqua. (2011a). Initial Field Survey, December 2011. PEPAM-Aqua, Kaolack, Senegal.

PEPAM-Aqua. (2011a). Start-Up Study of the Project to Support the Improvement of Water Quality in the Kaolack, Kaffrine, Fatick and Diourbel Regions, Final report, December 2011, PEPAM-Aqua, Kaolack, Senegal, 66p.

Sene S. (2016). Integrated Project of Sustainable Management of Rain Water with Multiple uses, Final report, March 2016, 12p. Horizont3000, internal document.

Sene S. (January 2019). Sustainable management of rainwater for multiple uses in rural areas, Final report, January 2019, 28p. Horizont3000, internal document.

Sene S. (2018). Integrated Project of Sustainable Management of Rain Water with Multiple uses, Semi annual report, July 2018, 14p. Horizont3000, internal document.

Chapter 11

Rainwater harvesting for improved food security and environmental conservation; Experiences from Malawi

Macpherson Nthara

Rainwater Harvesting Association of Malawi
macphersonnthara@yahoo.co.uk

Keywords: conservation, food, environmental, Malawi, rainwater harvesting, security, water

11.1 INTRODUCTION

The scarcity of water in different parts of the world has over the years forced people to find alternative sources of water. Rainwater harvesting (RWH) is one such technology that has proven to be more viable to complement conventional methods of water supply. Agriculture remains the main source of livelihood in semi-arid areas of Africa and will continue to dominate for the foreseeable future. However, land degradation and climate change have negatively impacted water resources and this is has led to reduced agricultural productivity leading to food insecurity in a number countries including Malawi. Rainwater harvesting technologies are required in these areas to impart stability in crop production during sub-normal years. The techniques help to maximize soil-water availability to crops and hence optimize crop yield per unit of available soil water.

11.2 CONTEXT

11.2.1 Geographical location

Malawi is a landlocked country located in Southern Africa between latitudes 9°22′S and 17°03′S and longitudes 33°40′E and 35°55′E (Figure 11.1). The main rain

 International Rainwater Catchment Systems Experiences: Towards Water Security
Editor(s): José Arturo Gleason Espíndola, César Augusto Casiano Flores, Raul Pacheco-Vega and Margarita Rosa Pacheco Montes
doi: 10.2166/9781789060584_0141

bearing systems are the Inter-Tropical Convergence Zone (ITCZ) and the Congo Air Boundary. Annual rainfall ranges from 700 to 2400 mm with mean annual rainfall being 1180 mm. The country's population of 4.4 million in 1966, has quadrupled to 17.2 million as of 2017 and is still largely dependent on natural resources. By 2022, the population is projected to grow to 19.4million and may continue to exert adverse pressure on access to services and overall socio-economic development of the country (GOM, 2017).

Malawi is relatively well endowed with water resources with 20 per cent of its total area covered by surface water bodies. While there seem to be abundant water resources in the country, the distribution across the country is irregular and varies by season and year. Ninety percent of the runoff in rivers and streams occurs between December and June, and only 0.1% of this is estimated to be captured for later use (GOM, 2017). Malawi is categorized among water stressed countries with less than 1700 cubic meters of freshwater per capita. The country's economy is largely dependent on agriculture and water resources are therefore central to socio-economic development. Water consumption in the country has undergone considerable increase for both rural and urban areas due to demand for domestic (3%), industrial (10%) and agricultural use (85%) (Mloza-Banda *et al.*, 2006). The stressed water resources are further challenged by high population growth, climate change and variability, increased sedimentation in rivers, lakes and reservoirs due to catchment degradation.

The Malawi Government estimates that 65% of the population has access to potable water. However, due to poor maintenance of supply systems only 40% of the population is actually served with potable water. The most common type of water facility used in Malawi is an unprotected well or spring while the most popular safe sources of water is a borehole. It has been observed that in rural areas, families tend to rely on unprotected traditional water sources which often get polluted in the rainy season. On the other hand, Malawi has ample water resources which translated into 18 billion cubic meters per annum as surface runoff. As of the year 2010, the estimated domestic demand was 95 million cubic meters or about 0.5% of total surface runoff.

11.2.2 The Rainwater Harvesting Association of Malawi

Rainwater harvesting initiatives received a major boost in the country with the formation of the Rainwater Harvesting Association of Malawi (RHAM). The association was launched in 2005 in Lilongwe, as a development partner toward integrated water resources management initiative of the Government of Malawi (GoM) and other stakeholders. It comprises of a number of individuals and institutions that are involved and take interest in rainwater harvesting activities throughout the country. The Association is also an integral network member of the Southern and Eastern Africa Rainwater Harvesting Network (SearNet) and the Flood Based Livelihood Network (FBLN). The association has a constitution and

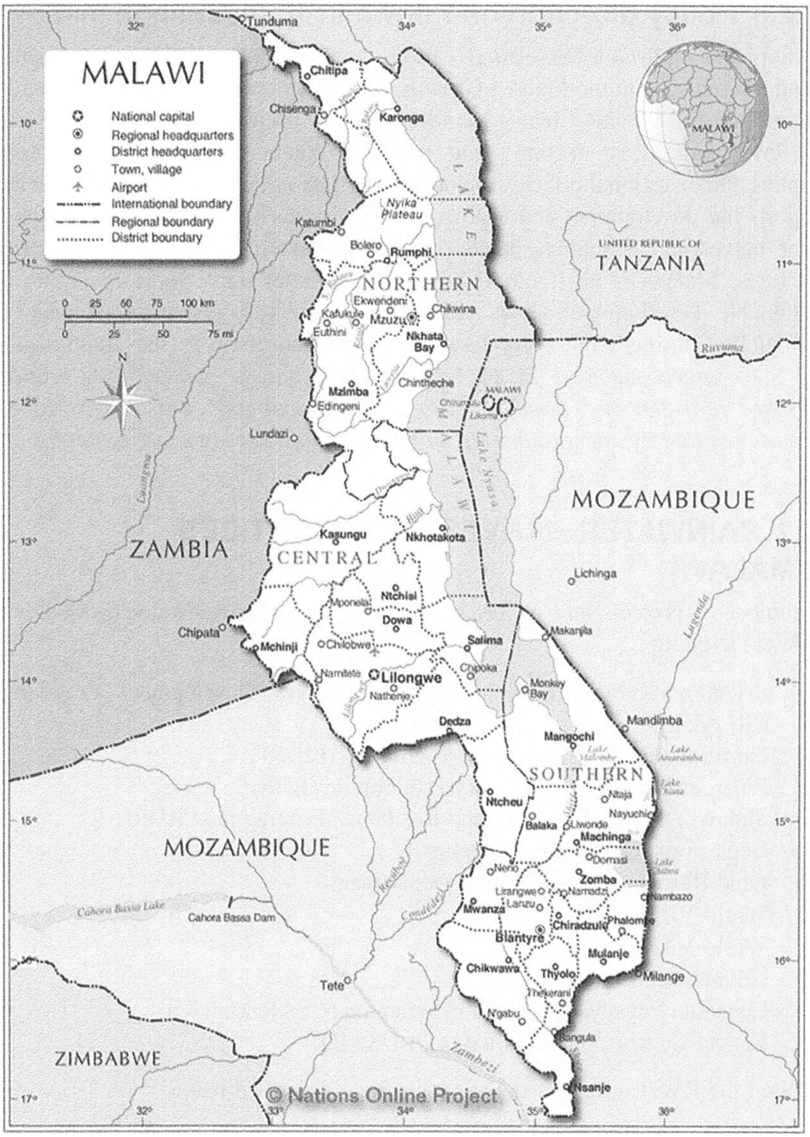

Figure 11.1 State of water resources in Malawi. (*Source*: Nations Online Project, Sajidu *et al.* (2013)).

is run by the executive committee through its secretariat. Since 2005, the association has collaborated with government and non-governmental agencies in organizing trainings, demonstrations, field days and a number of advocacy activities on rainwater harvesting across the country.

11.2.3 Policy direction on rainwater harvesting in Malawi

Promotion of rainwater harvesting is in line with the government developmental agenda as outlined in the Malawi Growth and Development Strategy III under the Agriculture and Climate Change Management key priority area. The MGDS calls for RWH as a way of mitigating water shortages and augmenting current supplies. Other national policies include the Irrigation Policy (GOM, 2017) which supports the development and testing of irrigation technologies including simple water harvesting techniques, hand dug wells, dams, weirs, and water control structures. Malawi is also committed to international frameworks including the Sustainable Development Goals (SDGs). For Malawi, provision of water to communities through rainwater harvesting will contribute to the attainment of two SDG targets, namely: Target No 2 – '*End hunger, achieve food security, improved nutrition and promote sustainable agriculture*'; and Target No 6 – '*Ensure availability and sustainable management of water and sanitation for all*'.

11.3 RAINWATER HARVESTING PRACTICES IN MALAWI

A number of projects and NGOs have been promoting rainwater harvesting in Malawi including:

- Irrigation Rural Livelihood and Agriculture Development Programme (IRLADP).
- Farm Income Diversification Programme (FIDP)
- Emergency Drought Recovery Programme (EDRP)
- Malawi Drought Recovery and Resilience Programme (MDRRP)
- Department of Irrigation Services
- Land Resources Conservation Department
- World Vision Malawi
- World Vision of Malawi
- Alliance One Tobacco
- Canadian Physicians for Aid Programme (CPAR-Malawi)
- Project Concern International (PCI-USAID)

Some of the RWH storage structures have been described below.

11.3.1 Above ground tanks

Above ground tanks for domestic application have been the most popular technique for rainwater harvesting in Malawi. The reasons for their popularity include: ease of colleting water from them; the water is usually of good quality; and the impacts they have on reducing drudgery of water collection over long distances. The first tanks were normally constructed using bricks and mortar (Figure 11.2). However, the last three years has seen the proliferation of ferrocement tanks and the use of

Figure 11.2 Above ground brick tank, Karonga District. (*Source*: Author's own).

Figure 11.3 Plastic tanks for rainwater harvesting, Balaka District. (*Source*: Author's own).

plastic tanks (Figures 11.3–11.5). To date over 500 above ground tanks have been constructed for roof-top water harvesting across the country.

The high cost of constructing these tanks remains a challenge to their wider adoption. On average, a 30,000 liter ferro-cement tank costs about 4000–5000 USD. As a result, most of these tanks are constructed with donor support. To upscale rooftop harvesting, low cost methods for water harvesting using the Calabash Cistern which cost about 300USD for a 5000 liter tank are being explored.

11.3.2 Lined underground tanks

Concrete lined underground tanks have mostly been built to collect runoff from ground surfaces e.g., school grounds, hill sides and road runoff (Figures 11.6 and Figures 11.7). The water from underground tanks is mostly used for cleaning, watering livestock and tree nurseries for afforestation. The Irrigation Rural Livelihood and Agriculture Development Programme facilitated the construction

Figure 11.4 30,000 L ferrocement tank Kasungu District. (*Source*: Author's own).

Figure 11.5 Ferrocement tank under construction. (*Source*: Author's own).

Figure 11.6 Underground tank, Karonga District. (*Source*: Author's own).

Figure 11.7 Underground tank (150,000 liters). (*Source*: Author's own).

of over 50 Underground tanks in the early 2000 which were used for agriculture production using Drip Irrigation. However, poor maintenance has rendered most of these underground tanks unusable. There has been lack of ownership of these tanks with communities expecting the donor to maintain the tanks for them.

Underground tanks have been popular in Malawi since they are able to collect huge volumes of water compared with the above ground tanks. The water is cooler making it suitable for irrigation. Their disadvantages include the large amount of sediment that is collected in the tank, and since the water is stored under the ground, withdrawing water cannot be done using gravity.

11.3.3 Dams

The history of dam construction for rainwater harvesting dates back to the mid-1940s when the colonial government formulated an irrigation policy with the primary objective of reinforcing water and soil conservation (Nthara *et al.*, 2008).

Figure 11.8 Water harvesting small dam, Senzani, Ntcheu. (*Source*: Author's own).

Figure 11.9 Mr. Mataka, a renowned water harvester in Ntcheu. (*Source*: Author's own).

Figure 11.10 Communally owned earth dam, Chitsime EPA, Lilongwe district. (*Source*: AMEI (2014)).

The colonial government constructed small dams as a conservation strategy and also as a source of water for the urban populace as well as for livestock Njoloma (2011) (Figure 11.10). Irrigation around small dams was an 'informal trend' that progressively developed either downstream or upstream of the dams. Between 1950 and 1980, the government of Malawi facilitated the construction of over 600 dams across the country (Figure 11.10). The dams were used for irrigation farming but some also supported livestock production and other activities. Due to land degradation, most of these dams are no longer functional as a result of siltation and lack of maintenance. Of late, there have been a number of projects that have constructed run off detention weirs for retaining surface run off (Figures 11.8 and 11.9). However siltation remains a major challenge for these structures requiring farmers to routinely de-silt them.

11.3.4 In-situ or soil storage rainwater harvesting

Farmers in Malawi are fully aware of the changes in climatic conditions evidenced by erratic rainfall and frequent dry spells. To address the situation, farmers practice rainwater harvesting in their fields in order so that they can harness any water that is available. Farmers have implemented a number of low cost in-situ water harvesting measures including swales, infiltration pits, percolation pond (Figures 11.11 and 11.12), conservation agriculture, box ridges, marker ridges, mulching and crop residue management, compost manure making and application, and the planting of grass strips/hedge-grows along the contours (Figures 11.13 and 11.14).

Figure 11.11 Infiltration pit, Neno District. (*Source*: Author's own).

Figure 11.12 Percolation pond, Neno District. (*Source*: Author's own).

In Liwonde Village, Neno District, rainwater stored in the ponds and hand dug wells is used productively in watering vegetables and fruit trees. Farmers are growing different vegetables (Chinese lettuce, mustard and tomatoes) in their fields that are recharged using in-situ techniques. The vegetables are growing well and are able to be used as a source income for the households. The success of in-situ techniques is attributed to the fact that they are cheap and easy to implement. Due to the huge potential of scaling up these in-situ techniques, more efforts have been devoted to capacity building of communities to enable effective implementation. The structures are continually monitored so that some minor maintenance is done by individual land owners. Implementation of in-situ techniques is achieved in clusters formed by the farmers. Choice of sites for

Figure 11.13 Check-dams, Karonga District. (*Source*: Author's own).

Figure 11.14 Infiltration pits, Karonga District. (*Source*: Author's own).

in-situ technologies is based on sites that have serious land degradation and can show the best results within a short time.

11.3.5 Flood-based farming systems

Flood-based farming (FBF) constitutes a highly productive, yet undervalued option for communities in flood plains to better manage flood water for crop and livestock production. Use of floodwater for productive uses is common in Chikwawa, Nsanje, Zomba, Nkhatabay, Salima and Karonga Districts. The following categories of flood-based farming are being practiced in Malawi:

(1) Flood-plain agriculture: cultivation of flood plains, using either receding or rising flood water. Farmers divert water into earth canals or concrete canals. Sometimes flooding rivers inundate adjacent field when they over topple their banks.
(2) Spate irrigation: diversion of short-term flood flows from seasonal rivers to field by means of small earth canals (Figure 11.16).
(3) Use of residual moisture: this the most widespread FBFS in Malawi especially in wetlands or Dambos (Figure 11.15). These areas are typically fertile, enabling farm households to intensively manage small plots of land with high returns for their labour and investment. As the moisture dries up, farmers will use water from shallow wells, using watering cans or treadle pumps to irrigate crops.
(4) Inundation canals: where cultivated fields are fed through a networks or canals by temporarily high water levels in perennial rivers.

These FBFS serve crop production, fishery, livestock, and are the sustenance of local ecological systems. Dependent on flood events, they are prone to climate change, yet they have considerable unused economic potential, as can be seen

Figure 11.15 Maize under residual moisture. (*Source*: Author's own).

Figure 11.16 Previously flooded fields planted with Maize. (*Source*: Author's own).

from the different experiences in countries in Africa and Asia. FBFS is, in essence, a resilience building block to smallholder climate change adaptation.

11.4 BENEFITS AND IMPACTS OF RAINWATER HARVESTING

Rainwater harvesting has been shown not only to improve the immediate water situation but also leads to a whole range of other benefits to the individual and communities that include: increased food and water security at household and community levels; provision of an appropriate water source at the point of use at low cost; provision of an independent back up water supply system supply for emergencies; development of employment opportunity and experience sharing skills for builders and artisans.

11.4.1 Lessons learned

Implementation of rainwater harvesting has demonstrated that there are opportunities to slow down land degradation and mitigate the impacts of climate change. Through integrated land and water management, both off and on-site ecosystem services can be derived that would lead to sustainable livelihoods. Rainwater harvesting can assist smallholder farmers to sustainably utilize low quality water to reduce pressure on high quality waters and preserve land. The following key lessons have been learnt from selected water harvesting projects.

Choice of farmers: In choosing farmers and sites for rainwater harvesting, future projects ought to focus on clusters of farmers who live and work in areas where there are serious water shortages and land degradation problems and which have a high potential to show positive results, rather than working with farmers regardless of their location in a watershed. It is important to target areas where there is greatest risk of land degradation from uncontrolled runoff.

Promoting few self-perpetuating practices: It will be important for rainwater harvesting projects to promote practices that will continue to spread among the farmers beyond the project period. While provision of inputs will be important, emphasis should be on practices that are of such benefit to a farmer that she or she will continue them independently and neighbors will emulate and continue to change the landscape. An example of such practice includes the excavation of swales and percolation ponds. Farmers use the residual moisture and stored water to grow vegetables, improving their income level and household food security (Figure 11.17).

Issues of incentives and subsidies: There is a general belief among extension staff that without incentives it would be impossible for rainwater harvesting practices to be taken up by a significant number of farmers. During sensitization meetings, it would be important for communities to realize the type of problems they are experiencing in their fields. Once the root problem is understood by the

Figure 11.17 A woman managing her vegetable garden, Balaka District. (*Source*: Nthara (2015)).

communities, no incentives or subsidies will be required for them to take remedial measures.

11.5 WAY FORWARD

There is need for a sustained campaign aimed at creating a mass movement on rainwater harvesting in the country. This will involve both the public and private sectors, NGOs, the media and the corporate world. At the heart of such a campaign will be school students who will be expected to carry the messages to their communities.

Due to high cost of some rainwater harvesting structures, there is a need to promote technologies that can be adopted at the small scale level. Much attention has so far been devoted to the promotion of in-situ techniques like conservation agriculture, manure making and application and Vetiver planting.

Capacity building in rainwater harvesting has to be intensified at all levels. This will include farmers, extension staff, and artisans.

There is need to lobby for investment in rainwater harvesting structures. While a number of NGOs have taken up the practice by constructing tanks, much remains to be done to upscale the practice in urban areas.

REFERENCES

AMEI (2014). Catchment Management Strategies for Linthipe catchment, Report No NWDP/IPC/021/09/010. Lilongwe: Ministry of Agriculture, Irrigation and Water Development).

Government of Malawi. (2017). Malawi Growth and development Strategy (2017–2022) – from Poverty to Prosperity. Government of Malawi, Lilongwe, Malawi.

Mloza-Banda H. R., Chikuni A. and Singa D. D. (2006). Small Scale Rainwater Harvesting for Combating Water Deprivation at Orphan Care Centres in Peri-Urban Areas of Lilongwe, Malawi. ATPS Working Paper Series No. 46, African Technology Policy Studies Network, P.O. Box 10081, 00100 General Post Office, Nairobi, Kenya.

Njoloma H. M. (2011). An Assessment and Inventory of Investments in Rainwater Harvesting and Small-scale Water Resources Development in Malawi: A Final Report. Rainwater Harvesting Association of Malawi (RHAM), Lilongwe, Malawi.

Nthara M. T., Manda M. J. and Mkwinda S. (2008). Technical Field Manual on Rainwater Harvesting, Land Resources Conservation Department, Ministry of Agriculture and Food Security, Lilongwe, Malawi.

Nthara M. T. (2015). Best Practices scaling up sites (SUPs) on Rainwater Harvesting and conservation agriculture, Rainwater Harvesting Association of Malawi, Lilongwe.

Sajidu S., Monjerezi D., Ngongoro C. and Namangale J. (2013). Malawi Country Water Resources Profile, University of Malawi.

Chapter 12

Challenges and opportunities in the implementation of rainwater barrels. An analysis of usability for the Guadalajara Metropolitan Area, México

V. L. Díaz[1] and G. E. Muñoz[2]

[1]Universidad de Guadalajara, Urban Projects Department, Guadalajara, México

[2]Universidad Autónoma de Guadalajara, Architecture Department, Guadalajara, México

Keywords: rainwater barrels, usability, user evaluation

12.1 INTRODUCTION

The aim of this chapter is to provide a description an evaluation of the usability of rainwater barrels in the context of the Guadalajara Metropolitan Area, in order to present the challenges and opportunities regarding the implementation of rainwater barrels. The analysis is based on the end-user perspective, of those who decide to implement the rainwater harvesting system in their households. In order to attain this, we implement a usability measuring instrument adapted to the evaluation of ecotechnological objects. This evaluation seeks to present an approach towards the identification of key elements for the improvement in the design of rainwater barrels. They are conceived as eco-products for the betterment of the environment and society as well as an affordable innovation.

Water is a fundamental resource linked to health and sanitation, and according to the United Nations (UN) water is an essential human right. The UN points out that there are approximately 884 million people without access to drinking water and 1.5 million children under 5 years of age who die from water-related diseases

Editor(s): José Arturo Gleason Espíndola, César Augusto Casiano Flores, Raul Pacheco-Vega and Margarita Rosa Pacheco Montes
doi: 10.2166/9781789060584_0153

(ONU, 2010, p.2). Hence, it is essential to integrate initiatives, programs and resources to assist the most vulnerable population. This situation encouraged the creation of UN-Water, a mechanism that focuses on the exchange of information, technology transfer and the creation of inter-institutional networks (ONU-Agua, 2009). Under this perspective, water is firmly established as a right and, in an economic context, as a resource. It is estimated that 80% of jobs worldwide depend on access to water (UNESCO, 2016, p.V), favoring in turn the economic development of countries. Investment in water is thus fundamental and establishes a correlation between the level of investment in water and the income of countries; in a global context where water is scarce, the capacity of its storage is linked to economic growth.

Irrespective of the fundamental role of water and the precedents set by the UN in terms of its care and management, there are national agendas that have failed to address the issue. In Mexico, while severe floods caused by rainy seasons directly affect highly vulnerable segments of the population, the issue of shortage and access to water remains unattended. In addition to that, a number of factors compromise the access to fresh water in the country, these include the pollution of rivers, lakes and seas on the one hand, and high consumption, negligent waste, disregard of the hydrological cycle balance and climate change on the other. Inadequate governmental water management and lack of infrastructure largely contribute to exacerbate this.

Within the ranking of the world's top 10 rivers at risk, one is located in Mexico, the Rio Grande, also called Rio Bravo (WWF, 2007). While 70% of the country's bodies of water are contaminated (Greenpeace, 2012, p.5), there is also a high extraction of water per person ratio of 691.40 m^3/hab/year, compared to China's 409.90 m^3/hab/year, Brazil's 306 m^3/hab/year and only 80 m^3/hab/year in Ethiopia (FAO, 2013). Moreover, the average municipal wastewater produced in 2017 was 7.41 hm^3 per year (234.9 m^3/s), while the municipal wastewater treated was 4.28 hm^3 per year (135.6 m^3/s) (CONAGUA, 2018, p.125). In 2015, the amount of renewable water per person was 3692 m^3 per year and it is estimated that by 2030 it will be reduced to 3250 m^3 per year (CONAGUA, 2016, p.79).

Considering that both the shortage of water suitable for human consumption and the excess of rainwater result in severe urban problems, attention to the management and consumption of water in its different variants of use is a high priority. In the case of the domestic use, there are some technical proposals to address this problem, such as the rainwater collection systems. According to the manual of rainwater collection of Texas, three stages define the process: water collection, conveyance and storage (Krishna *et al.*, 2005, p. 2). According to the United Nations Food and Agriculture Organization (FAO, 2013, p.9).

> '… Rainwater collection and use is understood as any type of technical effort, simple or complex, (…) meant to increase the amount of rainwater that is stored in the ground or in built structures, in such a way that it can later be used under conditions of water shortage'.

Although rainwater harvesting is not a new practice, going back to ancestral times all over the world, it is accepted that technological advances, combined with the idea of progress that supplied running water through complex hydraulic networks, were degrading this cultural practice to the point of disappearance. This can partially explain the reluctance of some society sectors to embrace this practice, since rainwater harvesting requires both an economic investment and a considerable change in habits. Currently, there are few commercial options in Mexico and a low perception of water vulnerability, these factors make the expansion of rainwater harvesting practices difficult.

This chapter takes elements from industrial design, evaluating the function of designed objects, in this case rainwater barrels. The benefits of this approach are described throughout this chapter, as well as the results of its application.

12.2 INDUSTRIAL DESIGN AS AN INTERVENTION TOOL

Industrial design is a fundamental tool to solve the problem of universal access to rainwater. Parsons (2015) states that, in order to achieve our objectives and perform our daily activities, we generate ideas, design and use our imagination to deal with specific problems or concrete needs, turning the act of designing into an underlying activity of life, one that is bound to all that man does. Papanek (1977), for his part, argues that through the conception of an object and, due to the impact it will generate in society, there is an unavoidable ethical and social responsibility on behalf of the designer towards humanity. Taking this into consideration, the conception of rainwater barrels demands their analysis as design objects which meet, through to their environmental, economic and social functions, the vital needs of man. In this regard, the notion of user-centered objects, a term and method developed by Donald A. Norman (2013) considers the position, characteristics, limitations and possible user responses as the center of decision-making within the entire design process, looking for products that are easy to use and comprehensible for the user through the identification of physical and cognitive interaction patterns while meeting the user's needs and wishes.

Within Nielsen's (1993) model of 'system acceptability', the perspective of the end user represents a basic condition to ensure the fulfillment of the interactivity functions of a product through the evaluation of a system or, in this case, a designed object. Centered on the needs and requirements of the user, Nielsen's diagram (Figure 12.1) branches off from the 'usefulness' attribute, which refers to the ability of the system to fulfill a desired objective, and is subdivided into 'utility', namely the system's ability to do what is needed based on its functionality, and 'usability', that is, how accessible that functionality is to the user.

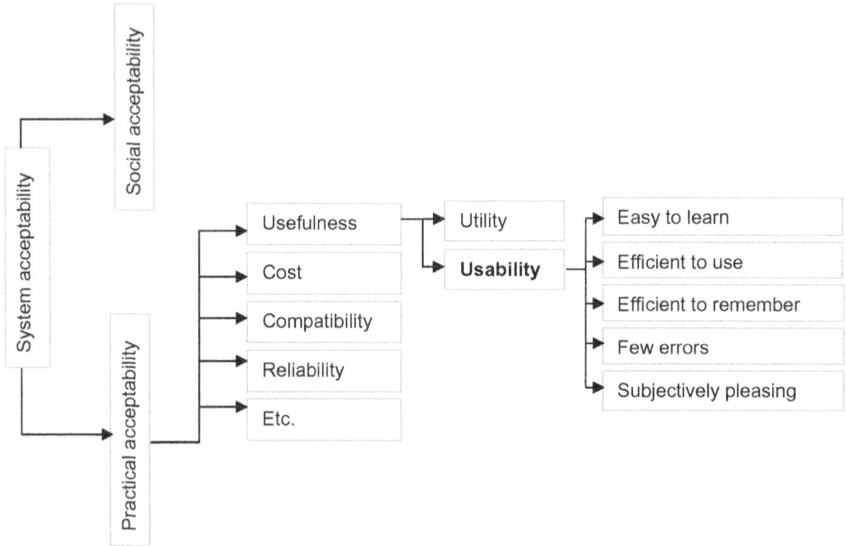

Figure 12.1 System Acceptability Model detailing the attributes with which 'usability' is assessed. (*Source*: Authors, adapted from Nielsen (1993)).

12.3 USABILITY AS AN EVALUATION TOOL OF DESIGNED OBJECTS

The assessment of usability as an evaluation tool can be traced back to the 1980s. It was, however, a decade later that Jacob Nielsen's (1993) works began to regard it as a substantial attribute that '(…) applies to all aspects of the system in which a human could interact, including installation and maintenance procedures' (Nielsen, 1993, p.25). As seen in Figure 12.1, this contemplates aspects such as ease of learning, efficiency of use, ease of remembrance, error tolerance and subjective satisfaction.

For the purpose of evaluating the usability of rainwater barrels, this work takes the four categories or techniques depicted by Nielsen and Molich's (1990) classification (Table 12.1) as point of reference, corresponding with the empirical type of evaluation for which further usability parameters were analyzed.

12.4 USABILITY MEASUREMENT PARAMETERS

According to the theoretical review carried out for this evaluation and, as condensed in a comparative revision of parameters (Table 12.2), it is evident that there are common variables in the measurement of usability. Among those 'ease of learning', 'attractiveness', 'efficiency', 'ease of use', 'persistence in memory', 'errors' and 'satisfaction' have served as the basis for the design of a usability evaluation instrument for rainwater collection barrels to be applied to our

Table 12.1 Types of evaluation

Formal evaluation	Assesses the user interface through some technical analysis. Formal analysis models are currently the object of extensive research applied primarily in software development projects.
Automatic evaluation	Computerized procedures for usability evaluation.
Empirical evaluation	Performs experimental user tests, with the aim of achieving a complete evaluation of the recipient. Currently most practical situations do not lead to empirical evaluations given the lack of time, specialization, inclination, or tradition to do so.
Heuristic evaluation	Performs the assessment of the user interface and generates a report according to the opinion itself (Ferrari & Mariño, 2014).

Source: Authors, adapted from (Nielsen & Molich, 1990).

Table 12.2 Usability measurement parameters comparison

Ferré-Grau (2001) and Alva (2005)	**Jakob Nielsen (1993, p.26)**	**Norma ISO 9241-11 (1998)**	**España and Pederiva (2012)**
Ease of learning	Learnability		Ease of learning
Understandability		Comfort and agreeability of use	Comprehensibility
Attractiveness			Measure of attraction
	Efficiency	Efficiency and effectiveness	Operativity
Remembrance in time	Memorability		
	Errors		
	Satisfaction	Satisfaction	Conformity

Source: Authors (2019).

case study in the municipality of Tonalá, Jalisco, a part of the Guadalajara Metropolitan Area.

Taking into account the application of usability parameters to the rainwater barrels' case study, some details are shown below. These contribute to their conceptualization as technological objects according to the categories of Ferré-Grau (2001) as well as the guide questions suggested by Jakob Nielsen (1993) and Nielsen (2012).

(a) Learnability: Ease of learning refers to the level of complexity represented by the use of the object, starting with the user's first contact, followed by the establishment of a self-generated knowledge peak caused by experience and interactive learning, up to the need for technical support. It addresses the question of 'How easy is it for users to accomplish basic tasks the first time they encounter the design?' (Nielsen, 2012).

(b) Memorability: Remembrance in time – this attribute considers the fact that users are not in permanent contact with the object in question, so that whenever a new interaction is necessary they are able to remember, without the need to learn 'from scratch', 'when users return to the design after a period of not using it, how easily can they reestablish proficiency?' (Nielsen, 2012).

(c) Understandability: This is based on the fact that users should be able to choose the product that best suits their needs and expectations, that is, to know the functions of the product and contrast them with what they need, in this case the input and output supply requirements of rainwater barrels. The language of the associated help documents must thus be simple, brief and in accordance with any potential user. The consideration of understandability is basic, so that false expectations among potential users are not raised. In its negative connotation, it would be linked to 'How many errors do users make, how severe are these errors, and how easily can they recover from the errors?' (Nielsen, 2012).

(d) Satisfaction: Attractiveness – it is generally accepted the level of visual attraction of objects implies an improvement in the display of information that enables interaction with the user and, to a certain extent, the preference of choice between several options. In reference to the perception of satisfaction, this would link to Ferrari and Mariño's (2014) question over how nice it is to use the designed object. Follow Nielsen, 'How pleasant is it to use the design?' (Nielsen, 2012).

(e) Efficiency: Once users have learned how to use the designed object, 'how quickly can they perform tasks?' (Nielsen, 2012).

From this perspective, all artificial objects have the ability to signify, communicate and relate to their users; in other words, the meaning itself is read in the 'language of the object'. In its broadest sense, that signification must act as container of the cultural and social assumptions (Sarabia, 1995, p. 114–117). The true challenge is thus to attend the broadest 'social, ethical and political problems (…) the artefacts produced by Design not only serve functional, symbolic and aesthetic aims, but also play a more fundamental role in influencing human life' (Parsons, 2015). On the other hand, Feng (2000, p. 164) as cited in Parsons (2015), argues that "even if technology is in some part driven by 'internal forces', (…) social values and demands also play a crucial, if not always highly visible, role".

In this regard, the scenarios turn into 'critical processes of learning and anticipation', in continuous adaptation, derived from the analysis of the possibilities of their impact (Sarabia, 1995, p 68). It is important to mention that the usability evaluation used for this work contemplates a systemic model comprising.

12.5 STAGES OF THE SYSTEMIC MODEL

The systemic model developed by Saravia (2006) represents the foundation from which the usability of the rainwater barrels' case study was evaluated. The six stages of development contemplated by the process are conceptually described in Figure 12.2, from this delimitation and analytical phases to the so-called 'Assessment Matrix' carried out on the last stage.

The usability evaluation instrument was applied to residents of the municipality of Tonalá. With the intention of establishing conceptual proposals that contribute to the improvement of the design object by addressing a specific problem, people were asked to collect water through the use of rainwater barrels in order to find out their degree of satisfaction and how they evaluate their operation. Stress was put on the stage of the process that includes their installation and operation, measuring their efficiency and effectiveness through the application of surveys and interviews.

12.6 RAIN BARRELS AND THE SCALL

Before delving into the discussion of rainwater barrels and their implications as rainwater collection systems, the extents of SCALL (Rainwater Collection System, for its acronym in English) should be first clarified and put into context. According to the 2005 Texas Manual on Rainwater Harvesting SCALL can be described as follows:

> 'Rainwater harvesting is the capture, diversion, and storage of rainwater for a number of different purposes including landscape irrigation, drinking and domestic use, aquifer recharge, and stormwater abatement. In a residential or small-scale application, rainwater harvesting can be as simple as channeling rain running off an unguttered roof to a planted landscape area via contoured landscape (…). More complex systems include gutters, pipes, storage tanks or cisterns, filtering, pump(s), and water treatment for potable use.' (Krishna *et al.*, 2005).

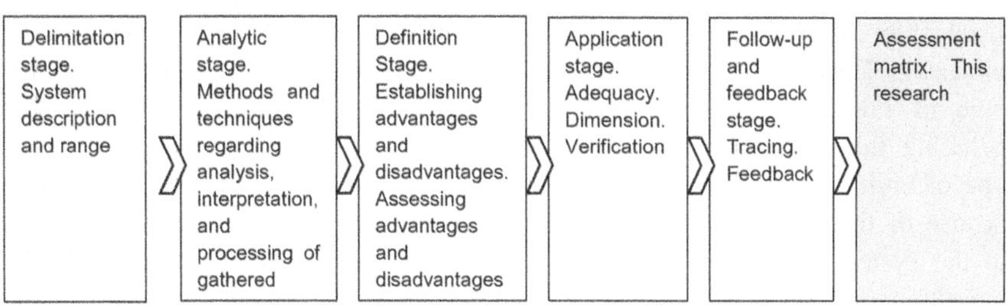

Figure 12.2 Systemic model applied to case study. (*Source*: Authors, adapted from Saravia (2006)).

The rain barrels case study represents an initial approach towards a SCALL (Rainwater call) that entails greater complexity on the technical (system maintenance capabilities, equipment monitoring, filter replacement, among others), economic (adjustments to the property, system parts acquisition, replacement of filters) and especially cultural aspects (change of habits, responsibility for cleaning the surface of the roof in contact with water, constant review of system parts, cleaning of containers, programming to change filters in a timely manner, among others) while considering the quantity and quality of water available for more crucial uses such as human consumption. In this context, rainwater barrels, according to the Institute of Technological Research of Water (IITAAC), are defined as:

> '… small containers used to capture rainwater from the roofs of buildings through gutters or downspouts. During rainfall, the drainpipes move the rain towards the barrel. If the container is full and it continues to rain, water that is not stored in the device comes out through a pipe located in the upper part of the barrel. This water should ideally purge on a green area away from the foundations of the house … ' (IITAAC, 2019).

The use of rain barrels aims to reduce water consumption from the public network, thus contributing to the mitigation of urban floods while diminishing the waste of rainwater that becomes normally polluted as it incorporates to the municipal network. Their use fosters, therefore, the development of a sustainable society where new generations are taught a different way of relating to their environments.

12.7 CHARACTERISTICS OF THE RAINWATER COLLECTION BARRELS

When the users decide to install a rainwater collection barrel, a product sheet is provided explaining the object's nature as a 'pilot program', letting the user know that it is an experimental initiative developed by IITAAC (Technological Water Research Institute, for its acronym in Spanish) Lic. Arturo Gleason Santana AC to promote environmental care and the encouragement of sustainable technologies among the population through the collection of rainwater. The sheet includes the following: (1) A brief introduction on the topic of rainwater barrels; (2) The location and characteristics of the site, including the dimensions of the collection surface and information about the type of building and water supply, among other data; (3) An explanation about the use of the rainwater barrel and its design (parts and adaptations that make up the system); (4) Water needs of the user/owner (s) of the barrel and the domestic (non-drinkable) uses of the harvested water; (5) Household water consumption figures and calculation; (6) Design adaptation proposal;

Figure 12.3 Rainwater barrel parts. A: opening, B: Cover, C: nozzle, D filter, E: purge. (*Source*: Authors, 2019).

(7) Installation process; (8) Installation cost estimate; (9 & 10) Designation of the person responsible for filing, reviewing and approving the document; and (11) Literature on the subject.

As shown in Figure 12.3, the rainwater barrel requires a collection surface, where gutters are placed to direct the flow of water through a pipeline, or downpipes, which flow into a filter in the opening of the barrel (A), which, in turn, contains a further filter (D) to prevent the entry of contaminants of significant dimensions, greater than the size of dust and sand. Once it has reached its maximum capacity, the excess water is purged (E), ideally redirected towards a gardened or soil area in order to avoid further waste; additionally, the cover allows for the cleaning of the barrel. Finally, the nozzle (C) near the bottom of the barrel provides access to the collected rainwater.

The designed object is a high-density and high-molecular-weight polyethylene container with adaptations made by technical specialists, although the barrel's capacity may vary, the one used for the purpose of this analysis features a 114 liter capacity. It can be installed in different types of buildings, however, use in domestic environments remains the highest priority. In general terms, its efficiency and relevance depend on the average annual rainfall of the Guadalajara Metropolitan Area; according to the Municipal Statistical Notebook of Guadalajara, issued by INEGI (National Institute of Statistics and Geography, for its acronym in Spanish), the average rainfall in the 1954–2001 period was 983.1 mm, while the driest year registered 615.2 mm and the rainiest year 1349.1 mm (INEGI, 2002, p.10). Roughly speaking, this translates into a great potential for rainwater collection in the region.

12.8 CHALLENGES AND OPPORTUNITIES IN THE IMPLEMENTATION OF RAINWATER BARRELS

The commercial availability of rain barrels as a product is relatively new, those evaluated throughout this study are therefore still in an experimental stage and access to users is limited, this, however, implies the opportunity for in-depth analyses by means of interviews. The instrument consisted of 35 questions related to the following categories: context, ease of learning, remembrance over time, understandability, attractiveness, efficiency (of use), satisfaction and errors. As mentioned earlier, every case-study barrel is located in the municipality of Tonalá.

Drawing on the results obtained, the use of the barrels appears to respond to a context of social relationships of trust, suggesting that the way in which the users became familiar with the designed object was through talks and recommendations by friends or acquaintances, establishing a base of trust regarding the rain barrels, which in turn encouraged the users to the acquisition and subsequent use of the product. Some stated that the most interesting aspect of the system is the self-created link to concepts such as 'ecology' and 'sustainability' and their latent interest in addressing these issues from the domestic sphere. Regarding the motive that triggered the transition from being interested in the subject to taking action, we found that, for most users, the economic aspect was key. According to the perception of users, the acquisition of the barrel represented a smart investment considering the high benefits expected in return; as a general rule, the obtainment of a container represents a first stage to then adjust, in a short to medium term, to a full system of rainwater collection.

As summarized, evaluation over usability concepts allowed us to establish a set of variables and measurement parameters which can be used for the assessment of the rain water barrels' case study. For the particular case of this work, we have considered to address the perspective of Saravia's (2006), the specific part of the process of his systemic model with an emphasis on its final stage, evaluated through the application of a survey and interview in order to understand the user's experience in regard to the operation of rain barrels with a focus on its qualitative aspects.

In terms of the **learnability** attribute, the system is generally regarded by users as one that allows the effortless collection of rainwater and is easy to install. However, concerns over the need for specialized technicians and adequate tools for this purpose were raised, noting that an improvised installation carried out by inexperienced hands might derive into problems that could confuse and discourage other people contemplating the purchase of such ecotechnology. Learning how to use it and how to perform simple maintenance actions was generally considered to be of low difficulty.

Being a utility that does not need frequent interaction with its user, in terms of **memorability in time**, details on the functioning of rain barrels could represent an obstacle for their correct operation. Likewise, problems would arise if constant preparation was required to be able to use the system or to keep it in good conditions. However, the users were able to report the characteristics, parts, and maintenance requirements of the system without requiring the re-reading of the manual that was delivered to them along with the rain barrel. Despite informing that they had to turn to the supplier for technical support, it was not on account of forgetting how to carry out an activity, but due to the requirements for adaptation to the installation.

In regard to **understandability**, as foreseen on the of ease of learning and remembrance over time attributes, the rain barrel system represents a simple-use ecotechnology, which is therefore easy to understand. Something important regarding this aspect is the impact that the dissemination of information exerts. As mentioned before, the socialization process that triggered the interest of the users, was sparked by recommendations of friends and by word of mouth. It is thus inferred that the person who recommended it in the first place understood the essence of the system, had formed expectations, and was subsequently able to explain and describe it before others.

The main expectations constructed around the usefulness of the rainwater barrel at the domestic scale lie on activities like garden irrigation, garage cleaning, laundry and vehicle washing. This is consistent with the expected level of water quality and it is assumed that its level of understandability is high. A further element that allows evaluation of this understanding is the instruction manual that was delivered to the users, observing that the high level of specificity and technical language used in it hinders and confuses the reader and might discourage its further use. This category is linked to the amount and type of errors in which users may incur as a result of an inadequate understanding of the system; since the errors reported by most of these users are limited to details regarding the installation process, a positive understandability of the overall process can be assumed.

The attribute of **attractiveness** reflects an overall poor perception; the rain barrel is considered unattractive and also not very adaptable to the aesthetics or character of the houses at which it was installed. However, when asked what they liked most about the rain barrel as a designed object, the answers had a utilitarian tone; the fact that it allows for the collection of rainwater is granted a greater value than aesthetics and attractiveness themselves. This reinforces the argument that barrels represent an initial approach towards the widespread of sustainable ecotechnologies that allows a gradual establishment and growth of the system, offering the opportunity to progressively tend to their aesthetic attributes.

In broad terms, levels of **satisfaction** among users point to their contentment in terms of the decision they took in order to carry out a concrete action to tackle the sustainability issue at a domestic level through the acquisition of the rain barrel. However, some details and inconsistencies between the expectations formed in the user and the perception of what has been obtained, tend to have a negative impact on their sense of satisfaction. Disappointment is perceived in regard to installation defects and inconveniences such as the lack of a stable base, or to the difficulty to extract water due to the position of the nozzle. Interestingly, the size of the rain water barrel is perceived as small, based on the initial expectation of the users derived from photos that they had seen before.

Considering that the **efficiency** category refers to the usability attribute and does not deal with performance nor specific tasks, the evaluation results reflect an acceptable level according to users, leaving room for improvement in terms of technical details. An example of this is the location of the nozzle, which makes it impossible to place a bucket under it. Likewise, given the high subsidies and low tariffs, the economic savings in regard to regular water expenditure are not considerable; within this context, rain barrels are not likely to be qualified as objects of efficient use.

The results obtained depict areas of opportunity where it is not difficult to adjust and improve technical aspects of the rain barrel and its installation process. These do not necessarily involve user interaction with ecotechnology and therefore do not reflect usability problems, but rather adjustments required by any technological object that goes through an experimental stage. Irrespective of this, some users mentioned that they would be willing to recommend the collection of rainwater through this product, as long as the technical aspects mentioned above are taken care of. Despite being an eye-catching, strange, colorful object, which is placed (mostly) on the front of their houses, users reported that barrels spark little or no curiosity among neighbors and relatives.

12.9 RECOMMENDATIONS

The main challenges and opportunities that were detected throughout the development of this study can be summarized as follows.

(1) Education acts a key element for the development of an environmental awareness that involves taking care of water and rainwater collection as sustainable and environmentally friendly alternatives;

(2) Adjustments to the existing Mexican legislation in terms of water protection are required in order to regard it as a living cycle to be respected and taken care of in the urban and domestic environments;

(3) Provide economic incentives to the people that implement the Rainwater Collection System in their homes or businesses while enforcing its application in diverse scales and variations for new housing developments in order to promote the protection of the water cycle in Mexico;

(4) Promote institutional programs that enable the dissemination of the benefits obtained through the existing rainwater collection barrel model and the creation of fiscal incentives for those who implement it; and

(5) Encourage and support the design and development of new affordable and scalable alternatives for rainwater collection while including them as part of public environmental policies to be prioritized by the government.

These suggest that rain barrels are an appropriate option to introduce society to issues of responsible water consumption, rain collection and other topics related to sustainability and environmental care, as they have a level of usability that allows users to upgrade to more complex systems. As in any experimental stage, the rain water barrel case study, as a system, requires certain adjustments that go from quality control of the installation and of its components, to revising the contents of the user manual in order to avoid hindering its comprehension. Figure 12.4 shows the location of the rain water barrel in front of the family house that implemented it in Tonalá.

It is worth noting that people who have acquired this product manifest their individual interest to address environmental problems, which in turn reveals the construction of a type of idiosyncrasy that is posited in favor of the environment and natural resources, one that has found an echo in initiatives such as this. Likewise, a broad dissemination of the pertinence and benefits of the domestic installation of Rainwater Collection Systems is needed, since this suggests the emergence of a segment within society that, despite having already developed the motivation to take action towards sustainability in the urban habitat, find a lack or scarcity of options in the market.

Assuming that the attribute of usability is indispensable, albeit not the only one, for an ecotechnological designed object to successfully find its niche in society, especially in the domestic sphere, the results obtained so far are positive showing that the rain barrels are perceived as an easy- to-use system of low complexity, inexpensive and beneficial to the environment.

Figure 12.4 Family using rain water barrel. (*Source*: Authors, 2019).

REFERENCES

Alva M. (2005). Metodología de medición y evaluación de la usabilidad de sitios web educativos. Universidad de Oviedo, Oviedo, Spain.

CONAGUA. (2016). *Numeragua México 2016*. México: CONAGUA.

CONAGUA. (2018). Estadísticas del agua en México. Government of Mexico, Mexico.

España S. and Pederiva I. (2012). Evaluación de la Usabilidad en un entorno de Arquitecturas Orientadas a Modelos. Valencia. Retrieved on February, 2019 from http://personales.upv.es/jopana/Files/Conferences/Evaluacion_de_la_usabilidad.pdf.

FAO. (2013). Captación y almacenamiento de agua de lluvia. Opciones técnicas para la agricultura familiar en América Latina y el Caribe. Santiago de Chile. Retrieved February, 2019 from http://www.fao.org/docrep/019/i3247s/i3247s.pdf and http://www.fao.org/nr/water/aquastat/maps/World-Map.ww.cap_esp.htm.

Feng, P. (2000) Rethinking technology, revitalizing ethics: overcoming barriers to ethical design. *Science and Engineering Ethics*, **6**, 207–220.

Ferrari S. and Mariño S. (2014). Guía de evaluación de la usabilidad para herramientas de minería de datos. No Solo Usabilidad. Revista Sobre Personas, Diseño y Tecnología, **13**, 1–10. Retrieved February, 2019 from http://www.nosolousabilidad.com/ articulos/usabilidad_mineria_datos.htm.

Ferré-Grau X. (2001). Principios Básicos de Usabilidad para Ingenieros Software. V Jornadas Ingeniería de Software y Bases de Datos (JISBD 2000), Madrid, Spain.

Greenpeace. (2012). Ríos Tóxicos. Greenpeace, México.

INEGI. (2002). Anuario Estadístico Jalisco. Aguascalientes. Government of Mexico, Mexico.

IITAAC. (2019). http://www.iitaac.org.mx/proyectos/barriles/.

ISO. (1998). Ergonomic requirements for office work with visual display terminals, ISO 9241-11. ISO, Geneva, Switzerland.

Krishna H., Brown C., Gerston J. and Colley S. (2005). The Texas Manual on Rainwater Harvesting. Texas Water Development Board (Vol. 1), Austin. Retrieved July, 2019 from http://scholar.google.com/scholar?hl=en&btnG=Search&q=intitle:The+Texas+Manual+on+Rainwater+Harvesting#0.

Nielsen J. (1993). Usability Engineering. Morgan Kaufmann, San Francisco, USA.

Nielsen J. (2012). Usability 101: Introduction to Usability, https://www.nngroup.com/articles/usability-101-introduction-to-usability/ (Retrieved February 21, 2019).

Nielsen J. and Molich R. (1990). Heuristic evaluation of user interfaces. CHI '90 Proceedings, (April), 249–256, http://doi.org/10.1145/97243.97281.

Norman D. (2013). The Design of Everyday Things. Basic Books, New Jersey, USA.

ONU-Agua. (2009). Guía sobre ONU-Agua. Retrieved February, 2019 from http://cenca.imta.mx/index.php?option=com_content&view=article&id=1844:guia-sobre-onu-agua&catid=89&Itemid=29.

ONU. (2010). Resolución aprobada por la Asamblea General el 28 de julio de 2010. Vol. **660**. ONU/United Nations, New York, USA.

Papanek V. (1977). Diseñar para el mundo real: ecología humana y cambio social. H. Blume Ediciones, Madrid, Spain.

Parsons G. (2015). The Philosophy of Design. Polity Press, Cambridge, UK.

Sarabia A. A. (1995). La teoría general de sistemas. Isdefe Ingenierìa de sistemas, Madrid, Spain.

Saravia M. H. (2006). Ergonomía de concepción. Su aplicación al diseño y otros procesos proyectuales: Editorial Pontificia Universidad Javeriana, Bogota, Colombia.

UNESCO. (2016). Informe de las Naciones Unidas sobre el Desarrollo de los Recursos Hídricos en el Mundo 2016: Agua y Empleo. Organización de las Naciones Unidas para la Educación, la Ciencia y la Cultura, París. Retrieved from http://unesdoc.unesco.org/images/0024/002441/244103s.pdf.

WWF. (2007). World's top 10 rivers at risk. World Wildlife Fund, Switzerland.

Section 4

Notable Technical Cases

Chapter 13

The success story of multi-purpose rainwater management system at Star City, Korea: Design, climate change adaptation potential and philosophy

Mooyoung Han

Seoul National University, Seoul, Korea

Keywords: climate change adaptation, Hong-ik-in-gan, multipurpose, rainwater management, Star City, win-win strategy

13.1 INTRODUCTION

Rainfall extremes such as floods and droughts are becoming more frequent or severe in many parts of the world due to climate change. As an adaptation strategy for coping with climate change, an ancient concept of rainwater harvesting and management is being revisited.

Star City, located in the north-eastern part of Seoul, near the Han River, South Korea is a major commercial/residential complex with four tall buildings, each having between 35 and 57 stories with more than 1300 apartments (Figure 13.1). The catchment area comprises 6200 m^2 of four rooftop areas and 45,000 m^2 of terraces and gardens throughout the complex. The Star City rainwater management system (RWMS) has been operating since 2007 until now (2019) and is receiving worldwide attention as a model water management system that supplements the existing centralized water infrastructure as a Climate Change Adaptation (CCA) strategy for floods and droughts (International Water Association [IWA], 2008) and as a model for water management regulation enforcement.

Editor(s): José Arturo Gleason Espíndola, César Augusto Casiano Flores, Raul Pacheco-Vega and Margarita Rosa Pacheco Montes
doi: 10.2166/9781789060584_0171

Figure 13.1 Location of star city in Korea. (*Source*: Author's own).

Han and Nguyen (2018) emphasized that the special features of Star City RWMS are: (1) the concept of multi-purpose systém; (2) the proactive management of flooding; and (3) the city government incentive program.

In this chapter, the design and operational details are briefly reported. The potential for climate change adaptation for flood mitigation is presented quantitatively. The philosophical meaning in water management will be addressed. And lessons learned from this project will be discussed for further promotion of RWHM.

13.2 DESIGN AND OPERATION OF STAR CITY RWHM

The design and operation data of Star City RWHM has been reported in many publications (Han & Mun, 2011; Han & Nguyen, 2018).

13.2.1 Design

Figure 13.2 is a schematic diagram of the Star City RWHS. A total of 3000 m^3 of water is stored in three separate tanks. The capacity of each tank is 1000 m^3 with the mission of flood mitigation, water saving and emergency supply.

The rainwater collected from these areas is piped to two separate water tanks located in one of the basement floors under Building B which is 35-storey apartment, which has a floor area of 1500 m^2. By making 2 m water depth, a total water storage of 3000 m^3 can be made. The rainfall runoff from the ground is routed to the first tank and the rainwater collected from the rooftops goes into the second tank. A course screen is used for pre-treatment, while within the tanks a J shaped calm inlet prevents resuspension of bottom sludge. Each tank is designed

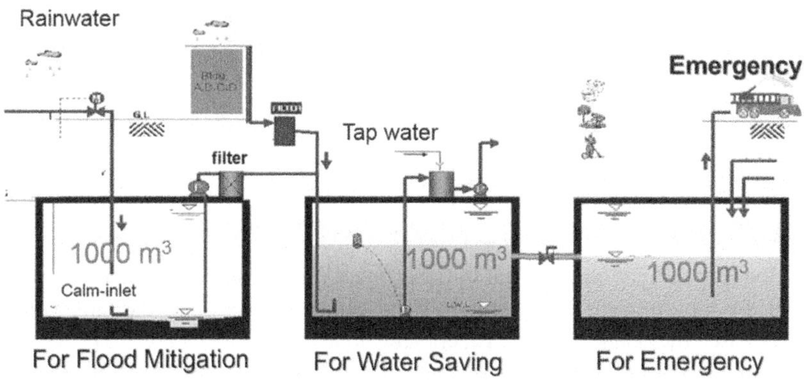

Figure 13.2 Schematic diagram of star city RWHM. (*Source*: Author's own).

to work as a plug flow reactor by installing baffles along the direction of flow, which can enhance the capacity of sedimentation. The water stored in the second tank (for water conservation, 1000 m^3) is used for irrigating the greened areas, flushing public toilets in the complex, and cleaning of roads and facilities. Garden irrigation is achieved efficiently by letting the overflow to the gutters connected to the second tank. When extreme rainfall is expected such as typhoon, the first tank (for flooding mitigation) will be emptied in advance before one day of arrival of typhoon to avoid the flooding. At this time, the sewer nearby can handle such a flow, because it is before the heavy rain arrives. It is designed to store the first 100 mm rainfall onto this site. The third tank (for emergency, 1000 m^3) is designed to store water for emergency at all times. It can be used as fire-fighting water or for emergency drinking and non-drinking water when the water supply is interrupted by any reason. Fresh tap water is maintained by regular replenishment of tap water after decanting half of the old water to the rainwater tank.

13.2.2 Operation data (Han & Nguyen, 2018)

13.2.2.1 Water quality

The pH of rainwater stored in the tanks is in the range of 6.2–8.5. The turbidity at the point of use is below 1.5 NTU due to the enhanced sedimentation in the plug flow type tank design and long detention time. There are no chemical parameters of concern because of the short mileage feature of rainwater. The water quality is good for irrigation and for toilet flush and cleaning without further treatment. During an emergency, it can be used for drinking by simple treatment facilities.

13.2.2.2 Water quantity

Water quantity can be normalized by using RUR (Rainwater Utilization Ratio) which is the ratio of the amount of rainwater utilized to the total annual rainfall fallen at the site. In most of the traditional design of land development, because the rainwater is designed to be drained, the RUR is near zero. However, based on the operating data for the first year, the volume of water that is saved is 26,000 m^3/year which is equivalent to RUR of 47%. By careful design and operation, higher RUR could have been achieved.

13.3 CLIMATE CHANGE ADAPTATION POTENTIAL

The flooding mitigation potential of Star City RWHM can be simulated by R-S-D model, by inputting the design rainfall and catchment area and the tank volume. In this case, the Tank volume/Catchment area is 5.8 m^2/100 m^3. The flooding mitigation potential is easily simulated by the method described in Han and Nguyen (2018) as in Figure 13.3. A 100-year frequency peak runoff when there is no storage tank of 26 m^3/h (point A) can be reduced to 18 m^3/h thanks to this rainwater tank (Point B). The peak runoff of 18 m^3/h indicates a 10-year frequency peak runoff (Point C). This means the downstream sewer which is designed for 10-year return period can be safe for a heavier rainfall of 100-year return period without increasing the sewer capacity. As such, the nearby area of Star City which was notorious as a flood prone area has not experienced flooding for last 15 years since the rainwater system was installed.

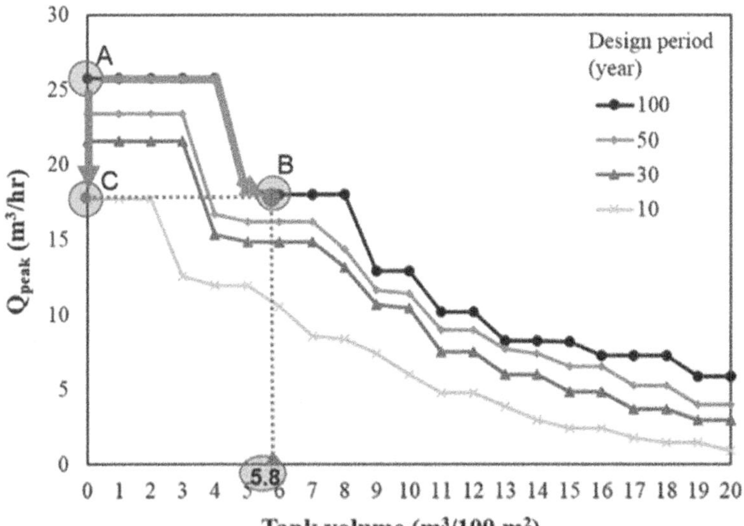

Figure 13.3 TP (Tank volume – Peak runoff) curves for R-S-D system using different design return period (using Seoul rainfall data and Huff method, normalized for 100 m^2 catchment area). (*Source*: Author's own).

13.4 PHILOSOPICAL CONSIDERATION OF STAR CITY

13.4.1 Win-Win process

The initial motivation of the Star City Rainwater System was for flood mitigation. Because the Star City site used to be a flood-prone area, the local government was afraid of further flooding risk after the development. They agreed with the construction company to build a rainwater system by giving financial incentive to allow building of 3% more floor area. After this successful demonstration project, Seoul city passed a regulation that can provide financial incentives when a developer makea a well-designed rainwater system. This Win-Win process made both public officials and developers happy. The tank design is also based on a proces to keep everybody happy. Among the three tanks of 1000 m^3 each, the first flood mitigation tank is for the benefit of downstream people, the second water conservation tank is for the benefit of residents, and the third emergency tank is made for the safety of all the nearby neighbors, which made everybody happy. This idea of everybody happy is based on the Korean 'Hong-ik-in-gan' Philosophy, which means 'make every party beneficial'.

13.4.2 Philosophy of 'Dong' (Village)

In Korea, each small village name ends up with Dong, such as Myoung-Dong and Insa-Dong, which has a special meaning for rainwater management. Star City RWHM adapted the concept of Korean 'Dong' philosophy. It means that water status should be the same before and after development using decentralized rainwater management (Figure 13.4). This is similar to the current rainwater regulations of advanced countries. For example, in Germany, they have a law to charge the sewer fee by the amount of increased runoff (2~3 Euro/m^2) after development. Or they are exempt the sewer fee when the owner makes a

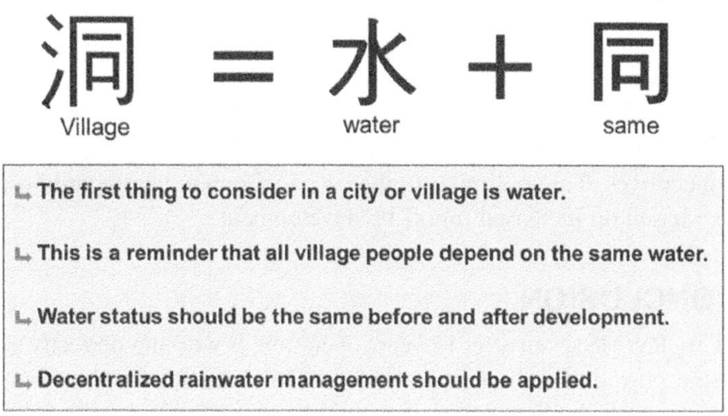

Figure 13.4 'Dong' philosophy for rainwater.

rainwater harvesting facility to cover the increased runoff. This is exactly the same as the Dong concept.

13.4.3 Five Commandments for water management

Han and Nguyen (2018) suggested the five commandments of water management which should be consider in urban water management. The star city can be a model for exactly following the rule.

Rain is money – Because the resident use rainwater to maintain their nice landscaping, the residents are enjoying a very cheap maintenance cost for the water usage in common area. According to the yearly bill for apartment, the water fee for the common use such as irrigation, cleaning and public toilet is only 20~50 cents per month for an apartment since the start of the Star City Complex in 2007. Therefore all the residents believes that the rainwater is money.

Distribute and manage rainwater – Star City site was formerly a baseball play ground of Konkuk University and used to be a flood-prone area. However, after the Star city rainwater management was constructed, there have been no flooding event for more than 15 years. Most of the rainwater that falls at the Star City site is not discharged to nearby sewer until the first 100 mm rainfall. So the rainwater at Star City can be managed easily.

Collect rainwater upstream on a small scale – The rainwater once it has fallen flows downstream by gravity. Because Star City collects rainwater at the point it falls, it is collecting rainwater at the furthest upstream point of the sewer systém and because it is not combined with other flow, the scale can be small.

Make it multi-purpose – Star City RWHM divided the rainwater tank into three tanks, with the missions of flood mitigation, water conservation and emergency. Sometimes, RWH system is thought not to be economically feasible because the water fee saved by rainwater is not very much. However, if it is designed for multiple benefits, the RWH system can be made economically viable and can make the society more resilient to climate change and social and natural accidents.

Social responsibility of rainwater management – It is logical for the developer to take responsibility for extra runoff created due to land development. In the Star City design, any rainwater up to 100 mm from a rainfall event that fell inside the site was collected and used as the responsibility of the building developer. The local government compensated the extra money for construction by providing some financial incentives. This is similar to the case in German law regarding charging a sewer fee based on increased runoff by development.

13.5 CONCLUSION

The Star City RWMS, even after 12 years operation, is working perfectly according to the design objective without any problem. The water fee for the common area usage is less than half a dollar per month per family. In the downstream neighbourhood which used to be a flood-prone area, there have been no flooding events reported. The emergency tank has never been used since the start of operations because there have been no accidental events to use it, however, it can still reduce the risk of accidents such as fire or sudden failure of water supply. The philosophy embedded in the design and operation of Star City RWMS stems from the Korean Philosophy of 'Hong-iK-in-gan', which is to make everybody happy. Based on this Win-Win system of offering financial incentives, more than 70 local governments in Korea have already made city ordinances to give financial incentives for RWHM systems. A new national ordinance is proposed to enforce the RWHM in Korea.

REFERENCES

Han M. Y. and Mun J. S. (2011). Operational data of the Star City rainwater harvesting system and its role as a climate change adaptation and a social influence. *Water Sci. Tech.*, **28** (11/12), 141–150.

Han M. Y. and Nguyen D. C. (2018). Hydrological Design of Multipurpose Micro-catchment Rainwater Management. IWA, London, UK.

International Water Association (IWA). (2008). Seoul's Star City: a rainwater harvesting benchmark for Korea. *Water*, **21**, 17–18.

Chapter 14

Developing a national rainwater harvesting standard

E. W. Bob Boulware, PE, M. B. A

Design-Aire Engineering, Indianapolis, Indiana 46141, USA
BBoulware@DAEngineering.com

Keywords: plumbing code, rain water catchment, standards, water, quality

14.1 INTRODUCTION

Considering the reality of diminishing availability of clean water, there are a number of reasons to consider rainwater harvesting and stormwater management as an alternative source for water: no water; bad water; stormwater management to reduce (prevent) flooding; reduce (prevent) combined sewer overflow (CSOs). By definition, rainwater is precipitation collected from an above ground, non-vehicular surface (the roof), and stormwater is recipitation that has come in contact with the ground (surface water).

While a rainwater collections system cannot 'make' water in the middle of a drought, or replace a dry well, the collection and management of rainwater serves to bridge the gap between rain events.

As wells are dug deeper, the quality of water can diminish, causing health issues for a community. Rivers and streams, too contaminated with heavy meals and harmful bacteria to be safely used, are not uncommon. Collecting rainwater as a stormwater management tool is growing in importance to diminish flooding and replenish ground water; and also serves to reduce overflows in combined storm and sanitation sewers.

Editor(s): José Arturo Gleason Espíndola, César Augusto Casiano Flores, Raul Pacheco-Vega
and Margarita Rosa Pacheco Montes
doi: 10.2166/9781789060584_0179

The benefits of harvesting rainwater are clear, but if rainwater catchment has been going on for millennia, why do we now need rules?

14.2 DISCUSSION

While rainwater collection for residential applications is common, when we expand applications to public buildings, more care is needed to manage the quality of the water. In a residential application, occupants become accustomed to their ambient waterborne bacteria. If we expand the applications to public buildings, more care is necessary to accommodate a lack of resistance to new ambient organisms.

Management of vectors such as vermin and mosquitos is also important. Unmanaged or improperly managed, rainwater collection can support water- and mosquito-borne disease. Examples of good intentions gone wrong occurred in rainwater collection systems built for New Orleans after Hurricane Katrina. Not enough attention was given to the water quality entering the cisterns, which was amplified by algae growth from improper screening of sunlight, served to make them better vehicles for mosquito breeding rather than potable water. As a result, rainwater collection, for a city built below sea level and surrounded by seawater, is not now allowed.

Aside from the safety aspects of having rules for safe rainwater collection, appropriate design and construction standards creates a level playing field for those designers and contractors. Without standards, a bad contractor can drive all the good contractors out of town. And as per the New Orleans experience, a bad job, where someone gets sick, can erase all the good jobs and create a negative impression of the rainwater collection industry.

The rainwater design standards developed in the United States (ARCSA/ASPE/ANSI Standards 63 and 78, the beneficial use of rainwater and stormwater, respectively) was an ARCSA collaboration to develop safe requirements suitable for application in public buildings. Rainwater harvesting standards in 2003 varied in the U.S. as cities attempted to address water shortages in their own way with no consistency. Some jurisdictions required chlorination of the water for toilet flushing, while others required no treatment at all. Some areas had objections that related to rainwater being perceived as a contaminated water source rather than the primary water source it is.

To obtain consistency, the best information available was gathered from research done in Australia, New Zealand, Britain, Germany and the United States. The primary objective was to standardize the inconsistent requirements that were prevalent at the time so it could universally be used safely. This included determining compatible system components and installation practices that would appease health and water utility officials.

The issues determined to be important in a rainwater collection system standard are: the rainwater collection surface and related water transport materials; the storage sizing and material; managing the water quality during storage and distribution; maintenance and periodic testing to assure ongoing water quality.

To be noted, in the United States there are important differences between a design guidelines, a standard, and a building code.

A Guide is a suggested design technique. It is an advisory, or a how-to design guide, that suggests design alternatives for multiple applications and recommended installation techniques. For example, design requirements for an above ground tank versus a below-ground tank, recommendations for sizing storage capacity, varying first flush volumes dependent upon site conditions, and water distribution options are included.

A Standard is a 'thou shalt' and 'thou shalt not' document. The emphasis is on protecting the public and not best design practices. It serves only to define the minimum passing grade for a system design and is intended to define requirements that will apply to all rainwater installations. A Standard serves as a reference document to a Building Code.

A Building Code starts with a Model Building or Plumbing Code that, in the United States, is provided by a national code writing body. The Model Codes are a collection of construction requirements that refer to many applications. Referring to the standards for each application (i.e. a rainwater collection system), rather than repeating or including the Standards, serves to reduce the size of the Code. The Code is further reduced as various jurisdictions review the Model Codes to modify, eliminate, or change for application in their respective jurisdiction. Only after the modified Model Code is agreed to and approved by local governmental bodies does it become a 'Building Code' and then have the force of law.

14.3 TECHNICAL

Standards for rainwater collection systems commonly focus on four factors: (1) the roof as a safe collection surface; (2) the quality of the storage tank; (3) the resultant quality of the water; (4) maintenance and testing to maintain the water quality.

14.3.1 The collection surface (roof)

The collection surface material needs to be non-toxic and not corrosive, which means that lead, ferrous, and biocide treated shingles are not allowed. Zinc (galvanized steel), is to be used carefully with an appropriate roof wash system to avoid health issues caused by high levels of zinc in the water. The roof material preferably should be a smooth surface, installed at an appropriate slope, in order to be self-cleaning, and with no overhanging trees for birds, animal residue, leaves· For flat roofs, there are NSF 61 compliant roofing membranes. NSF 61 (Formerly National Sanitation Foundation) defines the quality required for material in contact with potable water. Potable material requirements for rainwater collection systems were required, even for non-potable applications, so as to be available as a water source in an emergency situation.

The collection surface area is to be sized to compliment the storage capacity and the water need, with an appropriate safety factor applied consistent with alternative sources of water, i.e. availability of utility provided or a trucked-in water as an alternate supply. As a volume guide:

English: 640 gallons = 1″ of rain per 1000 ft^2,
Metric: 1 liter = 1 mm of rain per 1 m^2.

To transport the water, round bottom gutters are better than flat bottom gutters because they maintain a constant velocity at varying depths, which results in better cleaning of any sediment that might settle. To improve the water quality available for use, water entering the cistern should be properly filtered. If trees are adjacent, screened covers over the gutters are a good idea. As an alternative, a leaf filter may suffice to reject leaves and sticks from the inlet water (Figure 14.1).

A first flush device, that wastes the first part of a rain event, is beneficial to reduce the accumulated dirt and debris from the roof from entering the cistern. The First Flush Volume necessary to be wasted is dependent up on the site conditions and level of contaminants likely to be found on the roof (Figure 14.2).

Figure 14.1 Leaf Eater Debris Excluder. (Courtesy: Rain Harvest Systems).

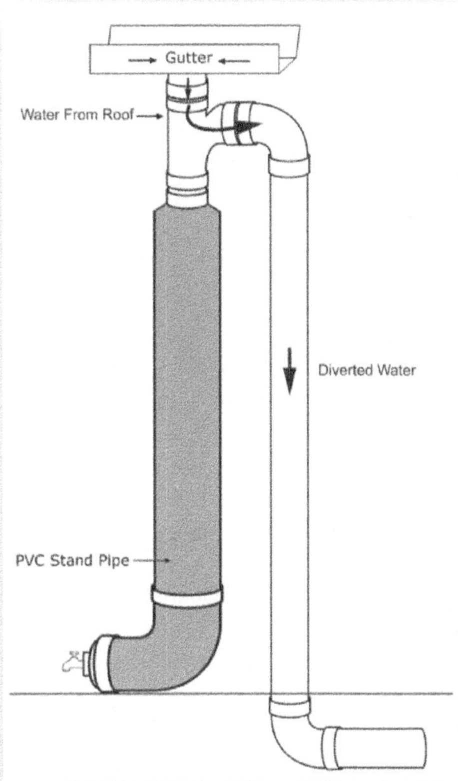

Figure 14.2 First flush device. (Texas A&M AgriLife Extension, s.f.).

For larger collection surfaces, the necessary water volume to provide a satisfactory roof 'first flush' can become large and unwieldy to be practical, making mechanical filters, such as provided by Graf and Wisy, a more appropriate means of cleaning the water entering the cistern (Figure 14.3).

These use mechanical features to allow debris to flow through the filter, thereby preventing the bulk of the debris from entering the tank, while pre-filtering to approximately 100 micron.

14.3.2 Water storage and distribution

Storage tanks, and the related distribution piping, need to be non-toxic and compatible with the storage of rainwater. The tank and related piping also need to be consistent for potable water applications. Since rainwater is similar to distilled water in regard to mineral deficiency, storage tanks need to be resistant to the

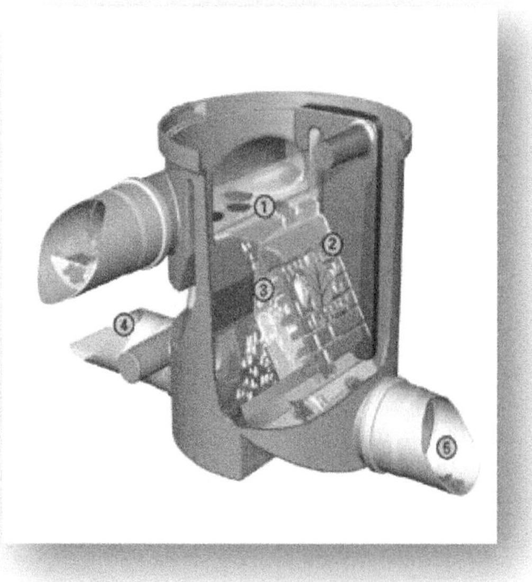

Figure 14.3 Mechanical roof wash device. (Courtesy: Graf Corporation).

slightly negative pH and the absorption of minerals. Suitable materials for rainwater storage are concrete and plastic.

Concrete as a tank material has advantages of being usually available locally which offsets shipping cost from a plastic tank. Leaching of lime from the concrete into the water serves to neutralize the pH of the water. Plastic, made from virgin plastic and not recycled plastic, is also a satisfactory material due to the durability and resistance to corrosion. Redwood has been popular as a water tank storage material but is now less popular due to the expense and environmental concerns of using a wood of limited sustainability.

Access ports and air vents for underground tanks should be a minimum of 4″ above grade to reduce possibility of surface water entering the tanks. Rainwater entering the storage tank should preferably have what is called a 'quiet inlet' (Figure 14.4). This is achieved by using a fitting to divert the water upward in the tank to avoid stirring up any settled debris at the bottom of the storage tank. All outlets should be properly screened for mosquito and vermin prevention. Above ground tanks should be located such that the sun will not heat the water, thereby providing an environment for algae and bacteria to grow in the water. And to assure water availability for emergencies, the tank and related piping should preferably be suitable for potable water contact (NSF 61 compliant). The overflow should be appropriately routed to prevent undermining of adjacent structures and tank supports.

The preference for sizing the tank storage volume is that it be sufficient for the intended use. For the sizing and design of a rainwater storage tank there are four variables for consideration:

- Water consumption needed – this can be approximated by calculating Monthly Demand;
- Monthly Demand = Volume/use × number of uses/day × 30 days/month;
- Time between rainfall events (to span dry season);
- Storage Capacity = Monthly Demand × time span (months).

This volume can be adjusted if an alternate water supply is available. In the US Virgin Islands 'Government Water' can be purchased if the water level runs out which would reduce the 'safety factor' needed in sizing the tank.

For underground tanks in areas of high water table, the upward force from the tank buoyancy needs to be considered to prevent the tank from floating when empty. The upward force (lift) is calculated as follows and can be offset by attaching the tank to 'dead men' or concrete ballast:

English: 62.4 lb/cubic feet.....8.33 lb/gallon;
Metric: 1 gram/cm^3............. 1000 kg/1 m^3.

Distribution piping is generally same as traditional plumbing, with the exception that piping needs to be compatible with the slightly acid nature of rainwater. This means that piping should not be ferrous or copper and polyvinyl chloride (PVC) should be used with caution due to leaching of chemicals into the water.

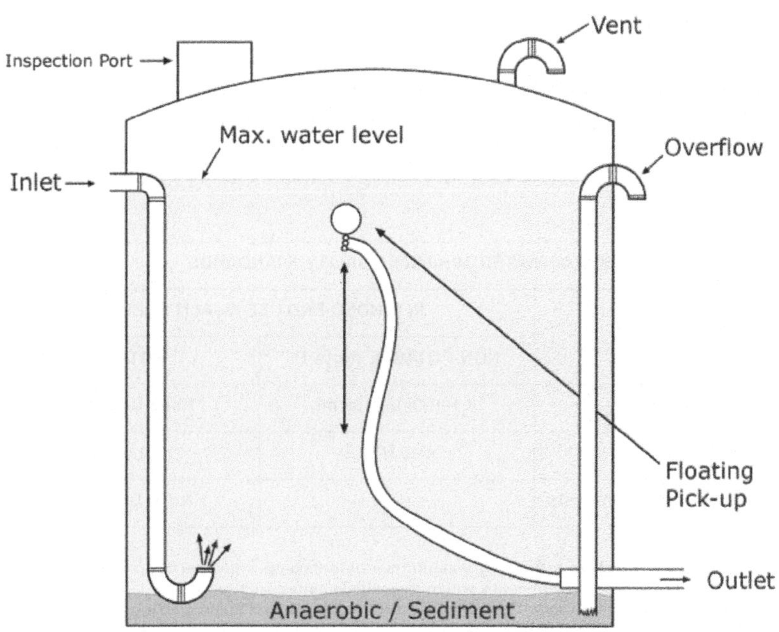

Figure 14.4 Typical rainwater storage tank piping with 'quiet inlet'. (*Source*: WC3 Design, 2013).

Cross-linked polyester (PEX) is preferred due to chemical resistance to leaching and being cheaper and easier to install. A floating water pickup in the tank, that draws the water below the water surface, where the cleanest water is, is preferred for potable water.

14.3.3 Water sanitation – maintaining water quality

Local authorities may have varying standards for water quality, but the American Rainwater Catchment Systems Association (ARCSA.org) recommends standards used by the Texas Rainwater Catchment Design Guidelines for potable and non-potable water applications (Figure 14.5).

Options for sanitizing water include chlorine, nano-silver particles, ozone, ultraviolet. chlorine, sand filter and thermal disinfection, all have been an effective disinfectant for bacterial disinfection. Hypo-chlorate tablets are commonly used although direct gas injection is also a possibility. Nano-Silver is a developing technology that has been effectively used in small applications.

Ozone is an effective disinfectant although it has no long-term residual in the water. Similarly, ultraviolet light does a good job inactivating microscopic, protozoan parasite such as cryptosporidium and giardia. Like ozone, UV has no residual value and should be located as close to the point of use as possible, downstream of all filtration. Sand filters have been used for centuries as a means of water disinfection and suitable for filtering common bacteria believed to be harmful. Heating rainwater to 60°C (140°F) will also kill most bacteria considered harmful.

Something to note regarding maintaining water quality is the importance of the biofilm that is created during storage. Dr. Peter Coombes (University of Newcastle, Australia) has done considerable research on the effect of biofilm and concludes that

STORED RAINWATER MINIMUM QUALITY STANDARDS		
PARAMETER	INTENDED END USE QUALITY LEVEL	
	NON-POTABLE (Note 1)	POTABLE
Fecal Coliform (e.Coli)	< 100 CFU / 100 ml	None Detectable
Turbidity	< 10 NTU	< 0.3 NTU
Viruses and Protozoan Cysts	----------------	None Detectible

Note:
1. Suitable for toilet and urinal flushing, washing machine makeup, and other approved applications in occupied space for environments with non health impaired occupants.
2. Monitoring requirements vary. Consult state and local guidelines for monitoring requirements.

Figure 14.5 Potable and non-potable water quality standard. (Courtesy: ARCSA).

Item	Roof	Tank Surface	Tank Outlet	HWS
Fecal Coliforms (CFUs/100 ml)	113	119	0	0
Total Coliforms (CFUs/100 ml)	310	834	127	0
HPC (CFUs/ml)	1318	3256	351	3
Pseudomonas (CFUs/100 ml)	49,825	6,768	4,433	0
Ammonia (mg/L)	0.39	0.1	0.11	0.18
Nitrate (mg/L)	0.25	0.06	<0.05	<0.05
Lead (mg/L)	0.015	<0.01	<0.01	<0.01
Zinc (mg/L)	0.55	0.25	0.17	0.15

Figure 14.6 Biofilm impact on water quality. (*Source:* Coombes, University of Newcastle, Australia).

Location or Criteria	FC (CFU/ 100 ml)	TC (CFU/ 100 ml)	Pseudomonas Spp. (CFU/ 100 ml)	Heterotrophic plate count (CFU/ml)
Surface	108	1,050	3,100	1,050
Mid depth	34	900	780	427
Bottom	55	862	4,060	1,252
Cold tap	<1	200	412	76
Hot tap	0	2	<1	<1
Guideline	0	0	NA	200

Figure 14.7 Bacteria count at varying tank water elevations. (*Source:* Coombes, University of Newcastle, Australia).

water coming out of a water storage tank can be cleaner than going into the tank (Figure 14.6). Furthermore, his research shows water in the middle of the tank has less bacteria than at the surface or bottom of the tank (Figure 14.7) which speaks to the benefits of obtaining water from below the water surface with a floating water pickup.

If water entering the tank is properly filtered with a first flush or other means, water quality in the tank, and the tank itself, can remain clean with minimal cleaning necessary. Keeping an above ground tank out of the sun to minimize tank water temperature is highly recommended to reduce the incidence of growing Legionella and other harmful bacteria in the tank. 32°C to 40°C (85°F to 105°F) is ideal breeding temperature for *Legionella* in a water storage tank, so care should be taken to keep stored water as cool as possible.

14.3.4 Maintenance and testing

For application of a rainwater collection system for a public building, it should be assumed that public scrutiny will be likely. Therefore periodic testing and proper record keeping of these tests is recommended.

After initial construction of a rainwater harvesting system, an 'initial commissioning' should be performed. This involves a thorough cleaning of all surfaces that come in contact with water. After a few rain events, the water should be tested to assure that the initial installation is capable of providing water to the intended level of quality.

At this point the system is turned over to the system owner/operator for his use and ongoing maintenance. Periodic water quality testing, to assure the collection surfaces and pre-filtration are properly maintained, with the test results retained to confirm the system is operating properly, is important. The frequency of testing may vary per jurisdiction, but testing potable water systems every three months, and non-potable yearly is a reasonable frequency of testing for most installations.

Cleaning of the tank is related to the quality of water entering the tank. If debris has been effectively removed from water entering the storage tank, the frequency of tank cleaning may only involve periodic inspection. If poor initial filtration is provided, a monthly cleaning may not be too frequent.

The above items represent issues that commonly need to be addressed by the Standard-writing Committee in developing a rainwater harvesting water standard. Once this is written, the next step is navigating the politics of getting the standard adopted as a public document.

14.4 POLITICS AND PUBLIC AWARENESS

Once the standard is established, it has no value if the public doesn't know about it.

Public Review (vetting) of the proposed standard is essential to gaining public acceptance of the proposed requirements and giving it credibility as a public document. Persons asked to review the document are those with a vested interest in rainwater collection and are being asked to review the document and voice their thoughts on proposed deletions, additions and changes. Those chosen to review the proposed standard would include designers, engineers, architects, contractors, city planners, related technical societies, health officials, and others interested in the safe implementation of rainwater collection systems.

The Standards Committee receives the written comments, reviews them, and votes whether to accept or deny the proposed changes. Because this is a public document, professional courtesy requires a public reply as to why the proposed modification was accepted or denied. 100% acceptance is not likely but a consensus appealing to the majority should be the goal and expectation.

Potential roadblocks to successful acceptance may include 'Old thinkers' (we've never done that before), public utilities (not wanting competition that will reduce sales), and competing special interests.

Once the standard is accepted by the public authorities, the education process of the public begins. Designers and engineers need to be taught about design methodology and architects need to be educated so that construction budgets can include a rain harvesting system. Building construction inspectors need to be familiarized so they know what to look for in a properly designed system. Building owners need to know about their systems so they are properly operated and maintained. And health officials need to be comfortable that safe water quality for the public will be maintained. All are necessary to be involved in the development, implementation, and acceptance of a rainwater collection standard.

14.5 SUMMARY

Developing a National Rainwater Harvesting Standard is a multi-step process for a public rainwater harvesting design standard (Building Code) to be effective. The technical issues that need to be addressed in a National Rainwater Harvesting Standard are: (1) collection from a safe (and approved) collection surface; (2) re-filtration to provide clean water into the tank; (3) quiet inlet so not to stir up debris; safe distribution of the water; and (5) appropriate filtration and sanitation of the water for delivery.

Once the standard is developed by the standard-writing committee, a constituency of stake holders need to review and approve the document to gain credibility before approaching 'Authority(s) Having Jurisdiction' for approval. When approval is obtained, the education of the public needs to begin to raise awareness about the rainwater collection systems, and the safe procedures necessary for their installation.

Once development and approval of the Rainwater Harvesting Standard is complete, those that participated can take pride in knowing that many people, now without a safe water supply, will be made better off from their efforts.

REFERENCES

Texas A&M AgriLife Extension. (n.f.). Rainwater harvesting. Recuperado el 15 de February de 2020, de https://rainwaterharvesting.tamu.edu/pre-storage-treatment/.

WC3 Design (23 de May de 2013). Recuperado el 15 de February de 2020, de ANATOMY OF A RAIN WATER HARVESTING TANK: https://wc3design.com/rain-water-harvesting-tank/.

Chapter 15

Stormwater management in transition in Brussels-Capital Region

Catalina C. Dobre

Faculty of Architecture La Cambre Horta, Unviersité Libre de Bruxelles, Place Eugène Flagey 19, 1050 Brussels, Belgium
cadobre@ulb.ac.be

Keywords: stormwater management, transition, Water Sensitive City, Brussels-Capital Region

15.1 INTRODUCTION

Worldwide, urban areas are being challenged to improve the stormwater management regime (i.e. the totality of beliefs, rules and practices that guide water management activities). Illustrated as the Water Sensitive City (WsC) by Australian researchers, the envisioned new regime aims to solve water problems, to adapt to future uncertainties (e.g. increase of extreme events), to create a more habitable urban environment and to reflect the aspiration of the community related to water (Brown *et al.*, 2009). Research indicates that a transition towards a new regime requires changes at different levels of society: a wider adoption of innovations at local level, a destabilisation of the current practices and an increase in external pressures forcing change (Schot & Geels, 2008).

Brussels-Capital Region (BCR), the capital of Belgium and of the European Union, is a representative case for the multilevel changes required to trigger a transition towards a WsC and for the key role stormwater management plays in this process. This chapter introduces four main characteristics of the stormwater

 International Rainwater Catchment Systems Experiences: Towards Water Security
Editor(s): José Arturo Gleason Espíndola, César Augusto Casiano Flores, Raul Pacheco-Vega and Margarita Rosa Pacheco Montes
doi: 10.2166/9781789060584_0191

management in the BCR that indicate an early stage of transition towards a WsC. Firstly, a change in perception of stormwater as a resource rather a nuisance is perceived in the modification of the water infrastructure and policies throughout time. Secondly, at present, water-related hazards, such as surface water pollution and urban flooding, together with environmental standards, are driving an urgent implementation of new actions in stormwater management. Thirdly, the fragmentation of responsibilities among state actors in the BCR's water sector often leads to failures in communication, but it sometimes enables new collaborations as well. Fourth and finally, in the BCR, civil society is particularly active in elaborating alternative actions by pressuring local and regional administrations to recognise them as relevant practices.

15.2 STORMWATER AS A SOURCE OF DAMAGE FOR THE URBAN ENVIRONMENT

In the middle of the 19th century, the rising rate of urbanisation increased the quantity of runoff. Stormwater, when combined with the Zenne River and its tributaries, became a constant nuisance in the urban environment. The Zenne watershed influenced in a large measure how rainwater infiltrates – favoured by sandy soils – and how it is drained easily on a flat topography. The eastern part of the BCR has a high declivity and a sandy soil, while the western part is relatively flat with a soil composed mostly of clay (Figure 15.1). Furthermore, the reduced space for the river due to urbanisation, and the lack of proper maintenance, led to several disastrous flooding events for the city (Deligne, 2003; Demey, 1990).

Inspired by similar initiatives in Paris and in London at the time, between 1867 and 1871, the Zenne River was covered in the central parts of the city with a large boulevard (Demey, 1990). It was not just the Zenne that received this type of treatment, but in parallel, or soon after, its tributaries were also covered. Covering the Zenne brought an important change in stormwater drainage. Underground pipes were placed alongside the culvert containing the river to capture stormwater and wastewater. This moment in the history of the BCR represented a change in the management of stormwater from a resource integrated in the urban environment to a source of damage at the same level as wastewater.

The model of a *tout-à-l'égout* (all to the sewer) infrastructure – to remove storm and wastewater as fast as possible in a combined underground sewer network – was well developed starting at the end of the 19th century in other large European cities, such as London and Paris. The underground sewer infrastructure was implemented in parallel with the water distribution network starting with 1850s (Yante, 2005). While the underground sewer system captures the excess of rainwater (runoff), the arrival of the water distribution network did not have an immediate impact on the use of rainwater as a resource for households. Rainwater harvesting was a common practice in the city for a much longer period. A study from the beginning of the 20th century reveals that up to 80% of the population were

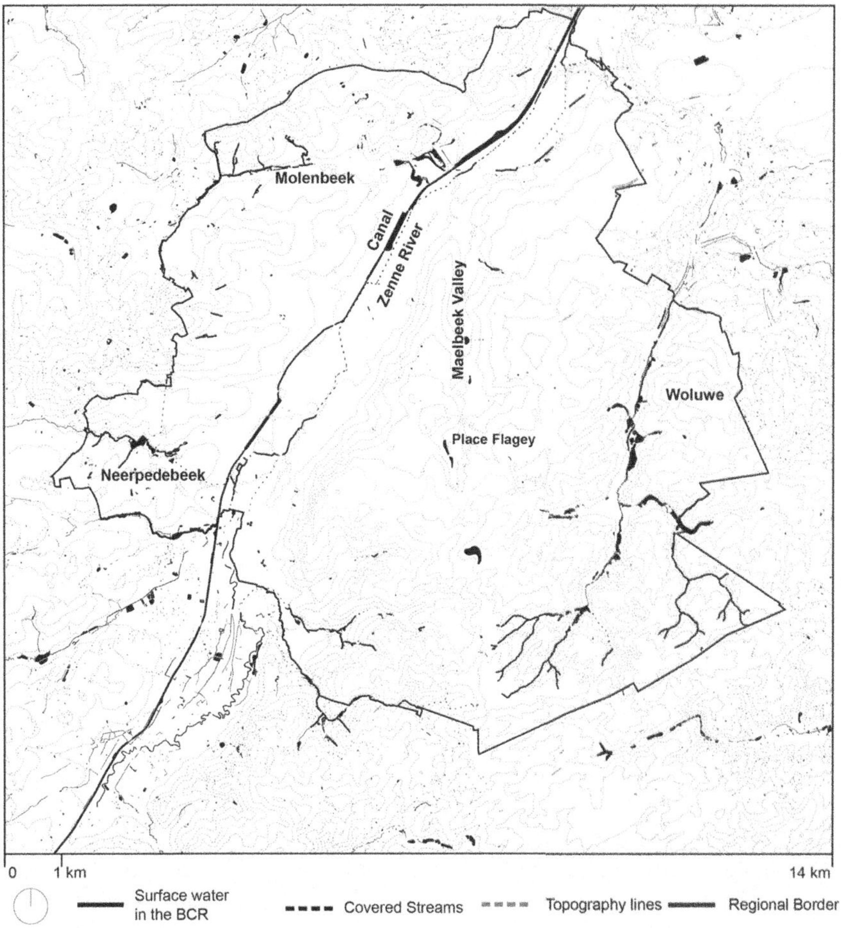

Figure 15.1 Map illustrating the urban streams and surface water part of the Zenne watershed located in the BCR. Map based on open source data from UrBIS BCR (topography, regional and municipal borders) and Bruxelles Environnement (watershed and hydrological network).

using rainwater tanks as an alternative to drinking water in the Brussels urban agglomeration and in the Flemish Region (Cornut, 1999).

Starting in the 1950s, the first signs of failure began to appear in the underground sewer system. The level of pollution in the Zenne River, the point of discharge of the runoff, became a cause of concern (Deligne, 2017). Moreover, between 1950 and 1978, four intense storm events are mentioned in a study carried out by the Ministre des Travaux Publics (1978). Among the mentioned causes were the expansion of impervious surfaces and the high quantity of runoff the sewer system needed to manage, together with an increase of wastewater, because of population growth. The study of the Ministre des Travaux Publics (1978) also points out a critical drop in quality of stormwater runoff as a result of industrial and urban activities. This study introduced the underground retention basin as an action to mitigate urban flooding by retaining stormwater during the pick of the storm and to release it in the sewer system afterword.

Urban flooding events and the pollution of the surface water network led to the laws of 1950 and 1967 that changed the ownership of surface water from private owners to the state administration to facilitate monitoring and dredging (Yante, 2005). The relatively slow implementation of surface water protection laws was due to the changes that occurred between the 1950s and 1990 in the administrative structure of Belgium. Together with the creation of the three regions – first Flanders and Wallonia, and later the BCR in 1989 – the administrative management of the water sector moved from the national to the regional level (Aubin & Varone, 2001). An important moment in the development of water institutions in the BCR also occurred in 1989. The creation of the *Institut Bruxellois pour la Gestion de l'Environnement* (IBGE) (Brussels's Institute for Environmental Management), today referred to as *Bruxelles Environnement* (BE) (Brussels Environmental Agency), followed the application of European Directives in environmental-related issues including surface water protection (Yante, 2005). Together with the creation of BE, reconsideration of actions to avoid surface water pollution began, such as the building of two treatment plants (Deligne, 2017). However, the actual implementation of actions took ten more years after the creation of BE (Aubin & Varone, 2001). Only in 2000, the BCR commission constructed a functioning wastewater treatment plant – South Station (approximately 20% of the wastewater) – and in 2007, the North Station completed the required capacity to treat all the wastewater in the region.

In parallel, the programme *Maillage Blue* (Blue network) initiated by BE started at the end of the 1990s, and since then, it has been included in all water and urban policies developed by the region. The programme aims to bring to daylight parts of the Zenne River and its tributaries and to ensure their flow by harvesting existing streams and springs. The connection of surface waterbodies has reduced the quantity of water the sewer system has to drain, and it has reduced the risk of flooding. Future water and urban policies influenced by the European Directive will integrate issues of surface water protection and public participation in close connection with stormwater management.

15.3 STORMWATER BOTH AS A RESOURCE AND AS A SOURCE OF DAMAGE IN WATER AND URBAN POLICIES

Influenced by the Earth summits on sustainable development (Rio de Janeiro in 1992 and New York in 1995), the European Water Framework Directive (WFD) came into force in October 2000, and it was designed to enhance the implementation of sustainable water management practices in the European Member States (Hering *et al.*, 2010). As it was located in the Schelde river basin, the BCR was required to follow the WFD guidelines and to release a river management plan in collaboration with other regional administrations located in the watershed – the Flemish Region, the Walloon Region, France, Germany and the Netherlands.

In 2006, the BCR's regional government agreed on delivering a *Plan de gestion de l'eau* (PGE) (Regional water management plan) (Bruxelles Environnement, 2015). Before the first PGE was finalised, in 2008, the region released a plan specifically for stormwater management – *Plan Pluie* (Stormwater Plan). Besides identifying the main causes of flooding in the region, the plan established a series of actions to diminish the risk of flooding (Bruxelles Environnement, 2008). According to the budget, conventional actions, such as the construction of underground retention basins or the maintenance of the existing combined sewer network, were the priorities. However, the plan supported a decrease in impervious surfaces by placing additional limitations on the design and construction of streets and buildings.

Eventually, the BCR's regional government adopted the first PGE in 2012, and it remained in force until 2015. The main objectives of the BCR's first PGE included the reduction of pollutants in surface and ground water, the qualitative restoration of the surface water network, the prevention of pluvial flooding and the integration of water into the urban environment (Bruxelles Environnement, 2015). The first PGE followed the principles of the WFD, and it was designed to minimise the impact of urban activities on the water cycle, considering that damage created by urbanisation to the surface water in the BCR can no longer be reversed (Mahaut, 2009). Visible changes were made to the structure and to the choice of measures implemented in the second PGE (2016–2021) (Bruxelles Environnement, 2015). For the first time in the programme, alternative stormwater management actions are recognised as complementary to conventional actions. Thus, the two PGEs recognised the different impacts of stormwater in the urban environment both as a resource – put in place through alternative actions – and as a nuisance, continuing to be managed under the approach of *tout-à-l'égout* in conventional actions.

While water policies have focused particularly on stormwater in relation to surface water protection and urban flooding, the urban regulations developed in parallel were designed to reintegrate stormwater as a resource into the built environment.

In 2018, the objectives included in the BCR's second PGE had few effects on urban regulations. In the RRU (regional urban regulations), there are mandatory regulations on stormwater management that are meant to follow the PGE in terms of water infiltration and retention: green roofs are obligatory for projects with a footprint larger than 100 m^2, water tanks of 33 l for every square metre of the roof surface and the maintenance of 50% of permeable areas (Brussels-Capital Region, 2019). Moreover, the current proposal to modify the RRU introduces mandatory regulations about the management of stormwater at the level of the parcel for new construction as well as for the extension of existing buildings (Brussels-Capital Region, 2019). At the municipal level, the Forest-Vorst local council delivered the first example of a plan of protection against flooding in 2012, and adopted a new RCU (municipal regulation) at the beginning of 2015. The RCU released by the Forest-Vorst municipality follows the obligations of the RRU with a more detailed explanation of stormwater infiltration and retention before it reaches the sewer system (Forest-Vorst Municipality, 2015).

15.4 PRESENT WATER-RELATED HAZARDS

15.4.1 Sewer system overflows as the main source of surface water pollution

Despite the construction of the wastewater treatment plants in 2007, the water quality of the Zenne River still presents a challenge for the BCR. As the natural flow of the river is small in comparison with the quantity of treated water that it receives, the level of pollution in the river is dependent on the functioning of the sewer infrastructure, including the wastewater treatment plants (Brion *et al.*, 2017).

A study carried out between 2009 and 2014 revealed how the pollution in the river is linked to the functioning of the sewer system (Brion *et al.*, 2017). During periods of heavy precipitation, both water treatment plants switch to the rainwater-operating mode: they only perform primary treatment to ensure they can treat a large volume of water. Furthermore, if the level of precipitation is very high, the water treatment plants completely stop functioning. The sewer infrastructure has a backup system to avoid overcharging the water treatment plants. There are 45 points of overflow, where the sewer infrastructure is linked directly to the river within the BCR (Bruxelles Environnement, 2015). The frequency of these overflows can reach 100 episodes per year in one single place (Le *et al.*, 2014).

15.4.2 Pluvial urban flooding

Overflow of the sewer system occurs not only into surface waters, but also into streets and household basements, creating the phenomenon of urban flooding. The drainage capacity of the underground sewer infrastructure is put to the test by the variable quantity of stormwater depending on the level of precipitation and the ability of the soil to infiltrate rainwater.

An analysis of the Royal Meteorological Institute on the precipitation level in the BCR between 1981 and 2010 revealed that most extreme storm events occur during the summer months (June, July and August), and that the annual precipitation averages 852.4 mm per square meter (relatively high for European cities, but similar to Amsterdam) (Tricot & Brouyaux, 2015). Nevertheless, scenarios for climate change in the region suggest possible changes in precipitation patterns:

during winter, many storms with large volumes of precipitation in short periods, and throughout summer, longer droughts that diminish the capacity of the soil to infiltrate rainwater (Tricot & Brouyaux, 2015). These studies indicate that soon, the combined sewer infrastructure of the city will need to drain more stormwater in shorter periods.

The quantity of stormwater to be drained also depends on the runoff, the excess of water that does not infiltrate into the soil. Research on the infiltration capacity of the soil in the BCR reveals that there are now fewer areas of the city that are not built up and that can allow infiltration (DeBondt & Claeys, 2010). Furthermore, research carried out in 1995 pointed out that 57% of the precipitation in the BCR area represents runoff (Verbanck, 1995). The results from 1995 are considerably different from the current situation, because the percentage of runoff increases with urbanisation, due to its larger impervious surfaces. In these conditions, the quantity of runoff the sewer system needs to manage increases exponentially.

Besides the level of precipitation and the quantity of runoff, there are other causes that exacerbate the risk of flooding, such as an aging sewer infrastructure, the disappearance of natural flooding areas and the fragmentation of the surface water network (Bruxelles Environnement, 2018).

15.4.3 Conventional actions to mitigate urban flooding

Developed in the 1970s, the underground retention basins are still the preferred action to mitigate urban flooding by the public administration. A study carried out by the Ministre des Travaux Publics (1978) pointed out the negative and positive aspects of implementing underground retention basins in the BCR. The high construction costs, problematic maintenance and technical difficulties in avoiding the formation of gas are critical points. The same problems of the underground retention basins are emphasised by more recent research. Mahaut (2009) considers them technical devices detached from urban reality. These devices are not easily accessible, they are larger than the human scale, they lack light and they do not follow the natural laws of gravity (they rely on electric pumps to empty them after the storm passes).

In parallel to conventional actions, alternatives have also been developed. Usually referred to in the French-speaking communities under the name of *mesures alternatives* (alternative actions) or *à la source* (at the source), they promote the infiltration, harvesting and use of stormwater as close to the place it falls as possible. Alternative actions, also known in Australia as water-sensitive urban design, use devices such as rainwater tanks, rainwater gardens or swales. Alternative actions are being developed in the region for new constructions or during the redesign of public parks. Referred to as a *Maillage Pluie* (Stormwater Network) by BE, these interventions remain punctual and are usually located in the low-density areas of the BCR.

15.5 FRAGMENTATION OF RESPONSIBILITIES BETWEEN REGIONAL, INTER-MUNICIPAL AND MUNICIPAL ADMINISTRATIVE LEVELS

The responsibilities in the water sector are highly fragmented between actors from the municipal, intermunicipal, regional and national level.

In relation to stormwater, non-profit organisations from the civil society intervene on the small scale in private spaces by using rainwater tanks or gutters to direct stormwater towards the street drains or sewer systems. The municipality manages the street drains, where the stormwater flows into the sewer network. Vivaqua oversees the small retention basins and the sewer network in which stormwater is harvested. The SBGE (Brussels's water management agency) manages underground retention basins larger than 5000 m^3, the main sewer collectors and the wastewater treatment plants. BE is responsible for the surface and ground water. All the actors managing the infrastructure are public entities at different levels of the administrative structure. SBGE and BE operate at the regional level, and they are directly linked to the regional government, while Vivaqua is an intermunicipal organisation.

The connection between the different levels of public administration is highly complex. The link between the regional and the municipal levels is made by each of the parliamentarians of the BCR, who also have a mandate at the municipal level (Delwit & Deschouwer, 2009). Intermunicipal organisations have similar types of structure. As an example, Vivaqua has an administrative board composed of municipal counsellors from each of the nineteen municipalities in the BCR and one mayor from a neighbouring municipality outside the region. At the municipal level, all the national, regional and language communities' legislation is coordinated. Representatives of municipal administration are crucial in the region, as they diminish the gap between the two levels of administration. Furthermore, the municipal administrative level has two main key values: proximity to the citizens, and operational efficiency in implementing trans-sectorial actions (Lagasse, 2012). However, the differences in total population, density of citizens are significant among the nineteen municipalities that compose the BCR (Delwit & Deschouwer, 2009). Thus, the capacity of each municipality to carry out all its functions is limited.

At the time of writing this chapter, no regulation stipulated which actors were in charge of stormwater management in the BCR (Bruxelles Environnement, 2018). The present complex structure of the BCR's public administration makes it almost impossible to identify one single actor in charge of stormwater management. Considering that stormwater is still mainly combined with wastewater in the sewer infrastructure, the lack of a separate responsibility for stormwater management does not come as a surprise. This should not be interpreted negatively. The lack of centralised coordination allowed the development of an integrated approach as opposed to the traditional sectorial approach in water management. The weaknesses of the fragmentation of responsibilities, such as inconsistency of the projects put in place, can be overpassed by a closer collaboration between actors. Moreover, with the integration of alternative actions in water policies and urban regulation, the fragmentation has allowed a wide range of actors to become involved, especially from civil society.

15.6 DYNAMISM OF THE CIVIL SOCIETY AS A LINK BETWEEN URBAN PLANNING AND STORMWATER MANAGEMENT

In parallel to the struggle to find the right balance among the state actors at the regional, intermunicipal and municipal level, civil society in the BCR has emerged as a means of counteracting the fragmentation and complexity of the institutional structure. To illustrate the impact of civil society in the water sector, a project in the public space, Place Flagey, is particularly illustrative.

Researchers consider the case of Place Flagey as a critical moment when urban planning met stormwater management in relation to the historical development of the Maelbeek Valley (the source of the Maelbeek River is currently entering the sewer system, but the valley is still visible in the topography of the territory and in the activities of the local associations) (Kohlbrenner, 2010) and to citizens' involvement in urban projects in the BCR (Houlstan-Hasaert, 2019).

In 1978, large flooding events were caused by sewer overflows in the nearby streets of Place Flagey (rue Gray) close to the centre of the BCR, towards the south-eastern part (Kohlbrenner, 2010). In the same year, a study carried out by the Ministre des Travaux Publics (1978) provided the first sets of actions to respond to the increase in flooding events: increase the discharge and storing capacity of the sewer system by the use of retention basins and to implement overflow systems so that the excess of water went directly into the surface water network. In the years to come, the local and regional municipalities often reflected upon the best approach to tackle the issue. A highway project in the Maelbeek Valley, together with the construction of a larger underground sewer system, was rejected and blocked by civil society, but a regional plan in 1991 to build a retention basin under the Ixelles ponds near the Place Flagey was better received by the inhabitants (Vantroyen, 1991). In 1996, the location of the retention basin was set to be under the Place Flagey, but the urban permit for construction was delivered only in October 2000 because of financial constraints at the regional level (Houlstan-Hasaert, 2019).

A collective composed of local inhabitants, Comité Flagey, considered the project as an outdated water management practice, without a clear understanding of its impacts on the neighbourhood's life (Comité Flagey, 2005). In response to the criticism of the project, the mayor of Ixelles mentioned in a newspaper article by Moyart (2002), after a storm event:

> This is quite catastrophic. The entire length of rue Gray was under 10 cm of water. (…) This is the first and true answer to those who say that the basin (the underground retention basin) is not useful.

Nevertheless, what was more important for the opposition was the way in which, on top of the retention basin, a new public square was going to be built. The administration considered an architectural call for projects a waste of time and money, and it proposed that the company assigned to build the retention basin should also design the public square. Inhabitants and NGOs expressed their anger towards the project, in part to the underground retention basin, but mostly to the lack of consultation on how the public square should be renovated.

> We had to accept the underground retention basin. We had to accept also the (underground) parking, the technical constraints (…) but why also the public square? (…) We had to propose alternatives while they (the regional and local administration) were proposing preposterous choices. (Comité Flagey, 2005, p. 24)

Urban activism in the BCR usually focused on opposing the view of the public administration. However, there was a perceived change in the case of Place Flagey, where the uptake of alternative actions evolved towards proposals for new solutions (Kohlbrenner, 2010). The collective Parcours Citoyen, active since 1997 in the Ixelles municipality, proposed public meetings to reflect upon alternatives to the underground retention basin (Houlstan-Hasaert, 2019). Furthermore, a new collective was created, Platforme Flagey, between neighbourhood committees, NGOs, a research centre and the nearby institute of architecture. The collective organised a call for projects that received intense public coverage and an impressive number of entries. The jury, composed of researchers, professionals, activists and inhabitants, analysed all of them and highlighted the important ideas that emerged for the future design of the square. During the proclamation of the jury, the head of the municipal council explained the struggles faced by the administration in relation to the project of the square:

> We were so busy with what was underneath (the retention basin) that we lost track of what was happening on the surface (public square Place Flagey). (Head of the municipal administration cited in Comité Flagey, 2005, p.27)

After this process, the regional administration took into account the conclusions of the jury, and it integrated them in a new international call for projects to select the architect to design the square (Houlstan-Hasaert, 2019). This process was also aimed at driving a change at the level of urban regulation, and at the same time repositioning architecture as a tool of interaction among various stakeholders, with particular attention given to citizen participation. Nevertheless, public participation in the urban revitalisation of Place Flagey ended there. Platforme Flagey transformed into a local point of information for the public without any ability to intervene in the selection of the architect (Kohlbrenner, 2010). The selected architects – Latz & Partner and D + A International, a collaboration between a German and a BCR-based landscape office – proposed a new design for the square that was built in four years and that was inaugurated in 2009 together with the underground retention basin (Figure 15.2).

For water management, the process in Place Flagey was a key point in starting to question whether underground retention basins are the most appropriate way to prevent flooding in the BCR. Inspired by the discussions surrounding the project in Place Flagey, for the first time in the BCR, a new concept *Nouvelles Rivières Urbaines* (NRU; new urban rivers) was developed, referring to alternative ways to manage stormwater on the surface to avoid flooding and to improve the urban environment (Mahaut, 2009).

Since 2009, the regional programme of constructing basins has continued, and in 2017, there were still six more basins to be implemented. Nevertheless, the case of Place Flagey received more attention than the construction of any other retention basin in the region. One of the impacts of this case was the increased involvement of the civil society in stormwater management. A new NGO, *Etats Généraux de l'Eau à Bruxelles* (EGEB), was formed in 2012 to reduce the gap between regional/local administration and inhabitants of water-related issues (Nalpas, 2011). Furthermore, what was particularly interesting and innovative for the BCR is that it was no longer possible to separate water infrastructure from the urban environment, even if it was constructed underground. More than that, the combination of water management and the redesign of public spaces opened up to new alternatives that moved the discussion from the table of engineers to the public, where multiple actors interact and have a specific expertise to share.

Figure 15.2 View of the Place Flagey in 2018 after the reorganisation of the square and the nearby lakes Ixelles. Underneath the square is located the retention basin partially used as a parking space. Photo credits: Costin Gheorghe.

Figure 15.3 Photo taken during a workshop organised during the *Ilot d'eau* design initiative. Photo credits: Meredith Dobbie.

The case of Place Flagey was just the start of several initiatives carried out by the civil society. As an example, in collaboration with the urban design office Latitude Platform and EGEB, in 2015, started in the *Forest-Vorst* municipality (southern part of the BCR) the *Ilot d'eau* design initiative funded by regional and municipal administration to facilitate public participation. The name *Ilot d'eau* (water building block) refers to the idea of a decentralised water system (*ilot* – translation in English – island) and to the traditional urban morphological unit closed by three or four streets (*ilot* – translation in English – building block).

The initiative was located in Saint-Denis neighbourhood the lower area of the municipality. The constant urban flooding events on the streets and basement (due to the overflow of the sewer system or the rise of groundwater) determined the creation in 1997 of the neighbourhood committee Stop Inondations. In this context, the *Ilot d'eau* design initiative focused on the co-planning, co-designing and co-construction of alternative actions (e.g., collective rainwater tanks) with citizens living in the same building block with the support of Stop Inondation (Figure 15.3). The practical results of the initiative – the construction of a common rainwater tank between two households and the creation of new green spaces at the limit between the facades and the public space – indicated the capacity of private spaces to contribute to an alternative management of stormwater. However, the struggles of realising the projects proposed by the inhabitants revealed the low capacity of the municipality to coordinate water-related project and the lack of economic incentives to implement common projects in private spaces (Dobre *et al.*, 2019).

15.7 LOOKING TOWARDS THE FUTURE

In the BCR, alternative actions are still far from being the dominant practice in the stormwater management regime. However, signs of transition towards managing stormwater as a resource on the surface, integrating of the community's values in relation to water and adapting to future uncertainties are already visible. At the local level, various local projects involving citizens have emerged. At the municipal level, diverse actors from civil society supported by urban design offices and academia have created an important momentum to facilitate the uptake of alternative actions in the stormwater management regime. Moreover, at the regional level, public actors, such as BE, and urban regulations for private spaces have opened up windows of opportunities.

As the two case studies illustrated, changes, seen as signs of transition, are temporal and localised. As previous research already indicated changes emerge randomly in specific contexts, rather than in a coordinated way for a compact transformation of the territory (Smith *et al.*, 2005). However, this does not mean that the changes do not influence one another. Actors, such as EGEB ensured the diffusion of new knowledge from the case Place Flagey to the *Ilot d'eau* design initiative. Questions remain about how both civil society and regime actors can collaborate to provide positive examples of how alternative actions can work in practice. Further research could look at their interaction in creating conditions favouring change in the way stormwater is managed in the BCR.

REFERENCES

Aubin D. and Varone F. (2001). *La gestion de l'eau en Belgique. Analyse historique des régimes institutionnels (1804–2001)*, (Water management in Belgium. Historical analysis of institutional regimes). *Courrier hebdomadaire du CRISP*, **26–27**(1731–1732), 5–75.

Brion N., Servais P., Garnier J., Kervyn M., Renders D., Evrard O., Lefèvre I., Ayrault S. and Deligne C. (2017). *Les changements de gestion des eaux usées domestiques et industrielles dans le bassin de la Senne: impacts sur la qualité des eaux* (The changes in the management of domestic and industrial wastewater in Zenne watershed: impacts on the water quality). In: Les rivières urbaines et leur pollution (Urban rivers and their pollution), L. Lestel and C. Carré (eds.), Quae, NSS-Dialogues, Paris, France, pp. 207–227.

Brown R. R., Keath N. and Wong T. (2009). Urban water management in cities: historical, current and future regimes. *Water Sci. Techn.*, **59**(5), 847–855.

Brussels-Capital Region (2019). *Reglement Regional d'Urbanisme. Titre I Caractheristiques des constructions et leurs abords* (Regional Urban Planning Regulations. Title I Characteristics of buildings and their surroundings), Brussels. [online] https://urbanisme.irisnet.be/actualites-accueil/enquete-publique-rru-2 (Accessed 10 September 2019).

Bruxelles Environnement (2008). *Plan Pluie* 2008–2011 (Stormwater plan), http://document.environnement.brussels/opac_css/elecfile/Plan_pluie_2008-2011_FR.PDF?langtype=2060 (Accessed 20 March 2018).

Bruxelles Environnement (2015). *Plan de gestion de l'eau* (River basin management plan RBMP). [online] http://www.environnement.brussels/thematiques/eau/plan-de-gestion-de-leau (Accessed 14 April 2015).

Bruxelles Environnement (2018). *L'évaluation préliminaire des risques d'inondation pour la Région de Bruxelles-Capitale* (Preliminary flood risk assessment for the Brussels-Capital Region), [online] http://document.environnement.brussels/opac_css/elecfile/RAP_20181218_floodriskassessment_FR.pdf. (Accessed 20 March 2018).

Comité Flagey (2005). *La ville pour qui, par qui ?* (The city of whom, by whom?). In: De la participaton urbaine (Urban public participation), J.-D. Bergilez, T. Bouhassoun, G. Grulois, and I. Lund (eds.), ISACF-La Cambre et La Lettre volée, Brussels, pp. 22–29.

Cornut P. (1999). *La circulation de l'eau potable en Belgique et à Bruxelles : enjeux sociaux de la gestion d'une ressource naturelle* (The circulation of drinking water in Belgium and Brussels: social issues in the management of a natural resource). PhD Thesis, Faculté des sciences, Université libre de Bruxelles.

DeBondt K. and Claeys P. (2010). Flooding Risk in Brussels, Water retention capacity of geological units 2p. In: We can change the weather, 100 cases of changeability, M. Wynants (ed.), VUB Press, Brussels, p. 223.

Deligne C. (2003). *Bruxelles et sa rivière. Genèse d'un territoire urbain* (12e – 18e siècle) [Brussels and its river. Genesis of an urban territory (12th–18th century)], Brussels, Brepols.

Deligne C. (2017). Les eaux de Bruxelles, systèmes et dépendances (Brussels's waters, systems and dependencies). In: *Les rivières urbaines et leur pollution* (Urban rivers and their pollution), L. Lestel and C. Carré (eds.), Quae, NSS-Dialogues, Paris, pp. 173–207.

Delwit P. and Deschouwer K. (2009). *Les institutions bruxelloises* (Brussels's institutions). *Brussels Studies*, 0–9.

Demey T. (1990). *Bruxelles. Chronique d'une capitale en chantier. Tome 1. De voûtement de la Senne à la jonction Nord-Midi* (Brussels. Chronicle of a capital under construction. Volume 1. Vaulting of the Senne at the North – South junction), Paul Legrain/Editions C.F.C., Bruxelles.

Dobre C. C., Ranzato M. and Moretto L. (2019). Citizen involvement in co-producing decentralised stormwater systems in Brussels. *CoDesign*, 1–18, http://doi.org/10.1080/15710882.2019.1631356

Forest-Vorst Municipality (2015). *Reglement communal d'urbanisme en matiere de gestion des eaux pluviale* (Municipal urban planning regulations for rainwater management), Forest Local Council, Belgium. http://www.forest.irisnet.be/fr/services-communaux/urbanisme-et-environnement/fichiers/rcu-eaux-pluviales-def-fr.pdf (Accessed 15 February 2018).

Hering D., Borja A., Carstensen J., Carvalho L., Elliott M., Feld C. K., Heiskanen A. S., Johnson R. K., Moe J., Pont D., Solheim A. L. and Van de Bund W. (2010). The European Water Framework Directive at the age of 10: a critical review of the achievements with recommendations for the future. *Sci. Total Environ.*, **408**(19) 4007–4019.

Houlstan-Hasaert R. (2019). *Le tournat esthétique de la participation urbain à l'éprouve de la société civile. Une recherche en terrains bruxellois* (The aesthetic turn of urban participation to the experience of civil society. A research in Brussels areas). PhD Thesis, Faculté d'architecture La Cambre Horta, Unviersité libre de Bruxelles, Belgium.

Kohlbrenner A. (2010). *Le chantier de la place Flagey: rupture ou continuité ? Approche historique des transformations dans la conception de l'espace public bruxellois (1958–2008)* (The Flagey Square construction site: rupture or continuity? Historical approach to transformations in the design of Brussels's public space). Master thesis, Unviersité libre de Bruxelles, Belgium.

Lagasse N. (2012). *Organisation des compétences entre Région et communes à Bruxelles: dépasser l'antagonisme des approches « Tina » et « Nimby »* (Organisation of competences between the Region and the municipalities in Brussels: overcoming the antagonism of the 'Tina' and 'Nimby' approaches). *Brussels Studies*, **61**, 0–12.

Le H. M., Petrovic D. and Verbanck M. (2014). The semi-sewer river: Hydraulic backwater effects and combined sewer overflow reverse flows in Central Brussels reduce deoxygenation impact further downstream. *Water Sci. Technol.*, **69**(4), 903–908.

Mahaut V. (2009). *L'eau et la ville, le temps de la réconciliation. Jardins d'orage et nouvelles rivières urbaines* (Water and the city, the time of reconciliation. Rainwater gardens and new urban rivers). PhD Thesis, Ecole polytechnique de Louvain, Université catholique de Louvain, Belgium.

Moyart J. (2002). *Flagey les pieds dans l'eau!* (Flagey with his feet in the water!) https://www.dhnet.be/archive/flagey-les-pieds-dans-l-eau-51b857b9e4b0de6db9a38520 (Accessed 15 February 2019).

Nalpas D. (2011). *Toutes les eaux de ruisseaux vont à la mer* (All stream waters go to the sea). Etudes et analyses/Inter-environnement Bruxelles. http://www.ieb.be/Toutes-les-eaux-de-ruisseaux-vont,4856 (Accessed 16 October 2018).

Schot J. and Geels F. W. (2008). Strategic niche management and sustainable innovation journeys: Theory, findings, research agenda, and policy. *Technol. Anal. Strateg. Manag.*, **20**(5), 537–554.

Smith A., Stirling A. and Berkhout F. (2005). The governance of sustainable socio-technical transitions. *Res. Policy*, **34**(10), 1491–1510.

Tricot C. and Brouyaux F. (2015). *2. Evolution du climat en région bruxelloise - température et précipitations* (2. Climate change in the Brussels region – temperature and precipitation), Bruxelles Environnement, Brussels, http://document.environnement.brussels/opac_css/elecfile/Clim 02 (Accessed 10 November 2019).

Vantroyen J.-C. (1991). *On creusera un bassin d'orage sous l'étang d'Ixelles la rue Gray aura les pieds secs (*A stormwater basin will be dug under the pond of Ixelles, rue Gray will have dry feets*)*. Le Soir. https://www.lesoir.be/art/on-creusera-un-bassin-d-orage-sous-l-etang-d-ixelles-la_t-19911112-Z04MDC.html (Accessed 7 October 2019).

Verbanck M. (1995). Transferts de la charge particulaire dans l'égout principal de la ville de Bruxelles (Transfers of particulate load to the main sewer of the city of Brussels). Faculté des sciences, Université libre de Bruxelles, Belgium.

Yante J.-M. (2005). *La gestion de l'eau en Belgique aux XIXe et XX siècles: Enjeux et acteurs* (Water management in Belgium in the 19th and 20th centuries: Issues and actors). In: The institutional arrangements for water management in the 19th and 20th centuries, J. C. Raadschelders (ed.), IOS Press, Amsterdam, pp. 81–94.

Chapter 16

Smart rainwater management and its impacts on drought resilience by Rural Semi-Arid communities: a case study of Northeast Brazil

J. Gnadlinger

Water and Climate Section – IRPAA – Regional Institute for Appropriate Small-Scale Agriculture, Juazeiro, Bahia State, Brazil
johanng@terra.com.br

Keywords: climate change, droughts, rainwater policies, resilience, sustainable livelihood

16.1 INTRODUCTION

There are two different approaches to understanding the semi-arid region of Brazil: the approach of the indigenous people and the approach of the colonizers. On the one hand, the indigenous population called the semi-arid region 'Caatinga' which means grey forest. This is due to the grey colour of the tree vegetation in the dry season. The region was a woodland with a green and a grey period depending of the season of the year. On the other hand, there is the colonizer approach. Portuguese colonizers arrived during the 16th century and began to deforest and burn the Caatinga to plant grass for cattle raising, transforming the region into a 'Drought Polygon'. In the 19th century, during the major drought of 1877 Emperor Peter II proposed to develop the region by the diversion of part of the São Francisco River, deep drilled wells (not appropriated in the mostly crystalline bedrock) and irrigation projects. These two approaches still determine the development projects of the region today.

Editor(s): José Arturo Gleason Espíndola, César Augusto Casiano Flores, Raul Pacheco-Vega and Margarita Rosa Pacheco Montes
doi: 10.2166/9781789060584_0207

Contrary to the colonizer approach, also in the second part of the 19th century, some Brazilian missionaries like Father José Ibiapina in the state of Paraiba, began harvesting rainwater, building in each Mission a 'water house', which was a rock water tank to store rainwater runoff, covered with a tile roof (thus 'water house') to improve the water quality and to diminish evaporation. Father Cicero Romão Batista encouraged in the rural area of the state of Ceará, the construction of rainwater cisterns by every house and of small stone barriers every 100 meters in intermittent streamlets and rivers to catch the rainwater runoff from the Caatinga to be used in the dry season. This chapter focuses on explaining how this understanding resulted in a new vision called 'Living in harmony with the semi-arid climate'. Different rainwater catchment systems technologies are presented as well as their implementation. They ensure safe drinking water for the families, water for the communities, water for agricultural production, water in emergency situations and water for the environment.

16.2 AREA OF REFERENCE

The semi-arid region of Brazil is located in the northeastern part of the country (Figures 16.1 and 16.2). In 2005, the Brazilian Government officially delimited the former drought polygon, calling it Semi-arid Brazil (SAB) (Brazil, 2005). SAB has a yearly rainfall below 800 mm, an aridity index of less than 0.5 and a drought risk of above 60%, calculated between 1970 and 1990. Due to the effects of environmental degradation, drought and climate change, the area of the SAB increased in the

Figures 16.1 Map of Brazil. (*Source*: Author, according to Brazil, 2005 and INSA, 2017).

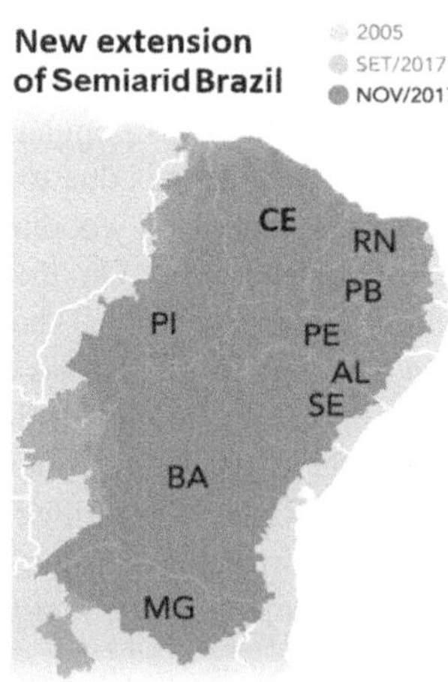

Figure 16.2 Semi-arid region of Brazil. (*Source*: Author, according to Brazil, 2005 and INSA, 2017).

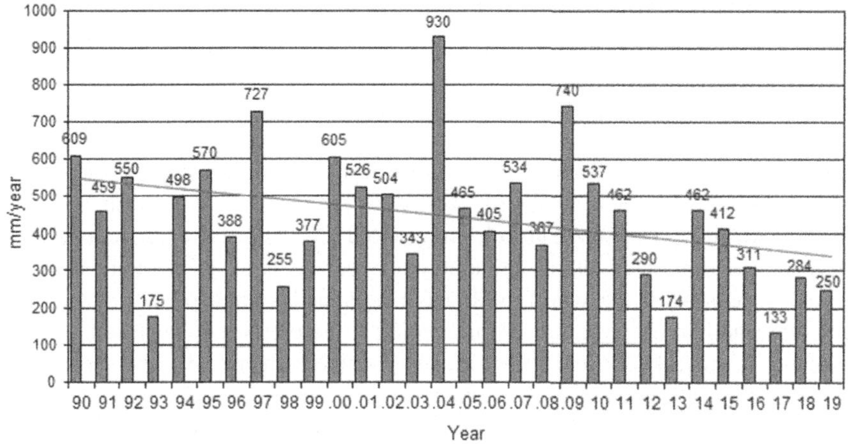

Figure 16.3 Annual rainfall in semi-arid Brazil, Juazeiro (1990–2019) with a trend line of decreasing rainfall. (*Source*: Author, according EMBRAPA, 2019).

last 15 years from 969,600 km^2 to 1,128,000 km^2 with 1262 municipalities, inhabited by 27 million people, 10 million of them live in rural areas (INSA, 2017). Traditionally the people make their living, collecting wood, honey, medical plants, processing wild fruits and raising goats and sheep without cutting down the native vegetation. In this region, the year has two seasons: a four-month rainy season and an eight-month dry season. It is particularly the irregular distributed rainfall in space and time, which characterizes the region. The city of Juazeiro in the center of SAB has an average yearly rainfall of 510 mm. In a drought year there may be only 133 mm rainfall, whereas in another year there could be 930 mm. Rainfall also has been decreasing over the last 30 years (Figure 16.3). Evaporation rate is high, due to continuous high temperatures (open surface evaporation of about 3000 mm a year). Longer droughts are part of this semi-arid climate and occur in a cycle of 25–30 years (Girardi & Girardi, 2001). The last severe drought event occurred between 2012 and 2017. In over 60% of the SAB, there is crystalline impermeable subsoil where it is not advantageous to drill wells. The normally small quantity of saline groundwater is not appropriate for human consumption or irrigation, although sheep and goats may consume it.

16.3 THE ANSWER IS TO LIVE WITH THE SEMI-ARID CLIMATE

When one, two or three years of low rainfall happen, it is not a catastrophe for the semi-arid region's plants and animals. Over a period of thousands of years, nature has been able to adapt itself to the semi-arid climate with its droughts and to build resilience. An event that drew attention was a drought between 1979 and

1983, which provoked immense internal migration, looting of governmental warehouses and which killed nearly up to one million people, especially the elderly and children (IBASE, 1989). While on one side, the corn crops withered in the fields and children starved to death filling the graveyards, the Caatinga region appeared to be beautiful, with their flowers, fruits and animals. Native plants of the Caatinga vegetation accumulate water and nutrient reserves, having tuberous roots and thick trunks to store water or deep roots to fetch it in the fissures of the crystalline subsoil; they avoid unnecessary evaporation; they produce and reproduce less in drought years, but do not die. Contrary to other semi-arid regions of the world, SAB has maintained a large part of its original tree, bush and shrub vegetation, which needs to be preserved and restored. This way the Caatinga keeps its resilience against climate change. The people have a popular saying: 'Preserve the Caatinga because it will resist climate change in the year 2100'. The lesson for the people is that their activities must also follow the concept of preparation for multiple drought years, similar to nature: they need to plan their water supply needs, not only for an eight-month dry season, but also for two years or more of drought.

Since the late 1970s, the Governmental Agricultural Research Center for the Semi-arid Region (EMBRAPA Semi-arid), conducts research on rainwater harvesting systems. In 1990, the Regional Institute of Small-scale Appropriate Agriculture (IRPAA) and other non-governmental organizations started undertaking research and dissemination of rainwater harvesting technologies, as part of the model 'Living in Harmony with the Semi-arid Climate'. During the decade it became necessary to create the institutional basis to implement larger programs, thus the government funded the Brazilian Rainwater Catchment and Management Association (ABCMAC) in July 1999. ABCMAC brings together researchers and users of rainwater technologies and above all, it organizes the biannual Rainwater Catchment and Management Symposium. In 1999, during the 9th International Conference of the Rainwater Catchment Systems Association (IRCSA), held in Petrolina, Pernambuco State, 50 experts from around the world presented experiences of rainwater catchment management. In particular, the successful experience from Gansu province, in the semi-arid region of China, became an example for SAB. In the same year, non-governmental organizations founded the Semiarid Network (ASA), which brought together more than 2000 grassroots organizations, including non-governmental organizations, farmers' unions, cooperatives, associations and church communities. Next, ASA launched the campaign under the slogan 'No family without safe drinking water' and proposed the One Million Cisterns Program (P1MC), to be implemented by civil society in a decentralized way (at the community, municipal, micro-regional, state and semi-arid regional levels). The P1MC was complemented by the Program One Piece of Land and Two Types of Waters (P1 + 2) meaning that every rural family must have: (1) a piece of land large enough to produce food, raise livestock, and ensure a sustainable life; and (2) two types of water storage, one for drinking and another for agricultural production (Gnadlinger, 2007). These two programs were important steps for Living in Harmony with the Semi-arid Climate, which also include other aspects such as agro-ecological food production, goats and sheep husbandry, marketing, contextualized education, political organization and environmental protection, needed to ensure sustainable development of the region. In this way, farmers, animal raisers, community representatives and even policy makers became aware how it is possible to live in harmony with the semi-arid climate and with droughts.

16.4 FIVE STEPS OF (RAIN) WATER MANAGEMENT

Experience has proved that despite the problems of uneven distribution of precipitation, high potential evaporation rates and unfavourable subsoil, it is always possible to catch water when it rains, store it and, therefore, have a safe source of water during the dry season, not only for drinking, but also for other uses (Gnadlinger, 2003, 2011, 2014). Rainwater management is part of the whole water cycle, which includes rainwater, surface water, water in the plants, soil water, groundwater and reuse of wastewater, even evapotranspiration. With infrastructure and good management, it is possible to have sufficient water for different uses in semi-arid communities, as well as during drought periods, especially if the following five steps are considered: (1) Water for families; (2) water for the communities; (3) water for agriculture; (4) water for emergency situations; and (5) water for the environment. These five aspects will be explained below.

16.4.1 Water for families

The most precious water is the one that is for human consumption, thus the supply of drinking water in case of scarcity has priority according to the Brazilian law (Brazil, 1997). For this reason, cistern water should be used first for drinking, cooking and basic hygiene. All families must have access to safe drinking water in quantity and quality. Drinking water should come preferably from cisterns, constructed near the house, large enough to store rainwater caught during the rainy season to be used during the long dry season (Heijnen, 2013). Several cistern designs are well known and used in the SAB. The two most cost-effective designs are: the semi-subsurface cistern made of pre-cast segments (which has user preference) (Figure 16.4) and the self-standing wire-mesh concrete cistern (for sustainability reasons) (Schistek & Gnadlinger, 2011; Thomas, 2001). Diversion of the first rain, removing water from the cistern with a hand pump and filtering the water before consumption guarantee safe drinking water. During the last 19 years around 1.2 million cisterns of 16,000 L for drinking water have been constructed by the organized civil society with funding from the federal government. As of March 1st, 2019, 619,943 cisterns had been constructed for families and 6848 cisterns of 30,000 L and 52,000 L for schools by ASA (2019). Figure 16.4 shows such a cistern, which especially benefitted women, liberating them from fetching water from long distances, and children, providing them with clean, healthy water.

Figure 16.4 One of the cisterns built by the ASA Semi-arid Network. (*Source*: Author).

From 2012 to 2017, the Ministry of National Integration also constructed *PVC (polyethylene plastic)* cisterns, manufactured by a multinational company, but criticized by the civil society because of technical quality problems and lack of community involvement. The P1MC methodology includes local participation in the construction process (hiring local masons, supporting the local economy with the acquisition of building materials, income generation, etc.) definitely advantageous when compared to the PVC cistern project where local participation is minimal, because the tank, the most expensive part, is produced in a factory (ASA, 2011; Heijnen, 2013).

16.4.2 Water for the Community

This water, which is for personal hygiene, washing of dishes and clothes, and for chickens, goats and sheep, may be provided by trench-like rock cisterns and by shallow wells. To guarantee this type of water, a community organization for planning, construction and maintenance is necessary.

Even in drought years, the 4-meter deep trench-like rock reservoirs with a small surface area to prevent evaporation have enough water for the needs of humans, livestock and a small vegetable garden. Figure 16.5 shows an example of this type of reservoir on April 15, 2012, when the picture was taken. The last rain had fallen in February (195 mm) and March (41 mm). The herd of goats had enough water for the following dry months. A similar situation occurred in 2016.

Figure 16.5 A four-meter deep trench-like rock cistern stores water for goats. (*Source*: Author).

To prevent evaporation and to increase the storage volume, one must not increase the width, but the depth and the length of the reservoir. The typical size is 5 meters wide, 4 meters deep and up to 30 meters long (with a storing capacity of 600,000 L). Obviously, it would be more appropriate to increase the reservoir volume by investing in greater depth and less length. However, the geological formation is a limit, because it is common that after four meters of excavation, hard impenetrable bedrock is found (Schistek, 2012).

Furthermore, more than 2000 manual water pumps with a flywheel (called Volanta pumps) were installed by ASA, especially in wells in crystalline subsoil with low output, providing water not only for sheep and goats, but also for other uses, if the water is not saline (Schistek, 2010).

16.4.3 Water for agriculture

Water for agricultural needs is supplied in many ways, including underground cisterns, sub-surface dams, ponds for salvation irrigation, road catchments, contour level plowing, rainwater stored *in situ* for fruit trees or planting crops adapted to dry climate conditions (sorghum, pigeon pea, green gram, sesame, etc.), use of manure and dry organic mulch to retain soil moisture for plants (Gnadlinger *et al.*, 2007).

Underground cisterns are promoted for multipurpose uses. The cisterns are cylindrical with a storage capacity of 52,000 L (Figure 16.6). Rainwater is collected from road catchments or runoff from 200 m^2 catchment areas constructed near the cisterns. These tanks permit supplemental irrigation of fruit trees and small vegetable beds (60 m^2), lined with PVC sheets 30 cm beneath the soil for avoiding water seepage. Until March 1st, 2019, ASA (2019) had constructed 103,528 cisterns of 52,000 L each to irrigate vegetables and fruit trees.

Sub-surface dams, appropriate in crystalline subsoil, store rainwater runoff: a transversal trench is dug (normally from 1.5 to 4 meters deep) perpendicular to an intermittent stream until reaching the impervious subsoil (Silva *et al.*, 2007). A subsurface dam in the SAB is composed of five parts: a catchment area, a storage area (or area for planting), a sub-surface dam, a shallow well and a spillway. The catchment area is the terrain represented by a small watershed where rainwater runoff from a stream or artificial drainage is collected. The storage or planting area is formed by the subsurface water table. When the water is stored in the soil, it is possible to plant all types of vegetables, corn, rice, beans or fruit trees. The dam itself is a perpendicular earthen wall with a PVC sheet placed on the downstream side of the trench, in a vertical position, from the impermeable crystalline rock layer until reaching the soil surface (Figure 16.7), intended to block the flow of surface and subsurface water upstream of the dam. The spillway has the function of eliminating the surplus water of the planting area after a heavy rainfall. A shallow well is dug at the deepest part of the storage area to use its water for livestock or small-scale irrigation.

Figure 16.6 Cistern of 52,000 L with catchment area for irrigation of fruit trees and vegetables, near the house a cistern of 16.000 L with drinking water. (*Source*: Author).

Figure 16.7 Sub-surface dam construction, placing the PVC sheet. (*Source*: Author).

16.4.4 Water for emergency situations

In years of drought, water for emergency situations is provided by deep wells and strategically distributed small and deep dams. This step is an interim solution when the three previous steps of water management are not enough. Over a period of thousands of years, nature has been able to adapt to droughts and build resilience. Supplying water by truck may be inevitable to mitigate the effects of drought, but this method is expensive, water often is of poor quality, and can be misused to make the underprivileged population dependent on politicians. The challenge is to make these programs unnecessary, when the first three steps are achieved. Drilling deep wells and constructing strategically positioned deep reservoirs, which do not dry out in drought years, can be one of the options for the population. However, in the SAB there are thousands of dams, which largely remain in the hands of an elite which do not share the water, leaving the surrounding population without access to that water, contrary to the first Article of the Brazilian Water Law (Brazil, 1997).

In late 2013, during evaluations about living with drought conditions, some examples of solutions were indicated: In the municipality of Palmas de Monte Alto, Bahia state, with an average rainfall of 800 mm, it rained only 300 mm in 2012 and 2013. However, it was enough rain to fill a cistern of 16,000 liters. This water was the best water for drinking, being used only for this purpose. Water provided by truck that came from polluted surface water sources was used only for livestock and vegetable gardens. In Itiúba, Bahia state, with 250 mm of rainfall in two drought years, the cisterns were filled by the water collected from the house roof, because the catchment area was greater than necessary to collect the water in one year of normal rain. It was suggested that every family should have more than one cistern. In this way, they could ensure drinking water with rainwater in drought time: in years of excessive rainfall, they could use water from cisterns also for other uses. In Sobradinho, Bahia state, in the community of Serra Verde, during 2013, from 200 mm of annual rainfall, 140 mm fell on April 22 and 23, 2013, enough to fill the trench-like rock cistern and have water for the livestock, until the next rainy season, which began in December 2013.

16.4.5 Managing water for the environment

Water management for the environment is based on the watershed, protection of springs and riparian vegetation, pollution prevention, wastewater recycling and reuse for irrigation purposes, supporting ecosystems and completing the water cycle of evaporation, condensation and precipitation.

Watershed Management is the integration of technologies within the boundaries of a drainage area for optimum development of land, water and plant resources to meet the basic needs of people and animals in a sustainable manner. The integrated management of small watersheds uses different rainwater harvesting technologies, starting from the highest point of the property/watershed, where the rainwater starts draining subsequently to the lower parts of the watershed.

On the higher parts of the watershed, work is done to promote eco-forestry and reforestation of Caatinga trees (re-caatingamento), plant groundcover plants tolerant to drought, vegetative barriers for soil, with natural pasture. On hillside areas, work should promote planting along contour lines, harvest water 'in situ', plant fruit trees and vegetable gardens, build small ponds for infiltration and groundwater recharge, recover gullies from the beginning of the water flow. At the lower parts of the watershed, it is best to use rainwater-harvesting technologies for food production, like subsurface dams, ponds, small gabion dams or successive barriers in streams to store water in the alluvium (See also Qiang & Li, 2009). A variety of watershed management programs in temporary rivers called hydro-environmental projects and natural vegetation recovery programs, called 're-caatingamento', are underway.

These five steps of water management make it possible to live with the semi-arid climate conditions and are the basis of the elaboration of decentralized and participative water plans, to be carried out in communities and municipalities of the SAB. They mean a paradigm shift in the management of water resources as opposed to large technical projects for the formerly so-called 'Drought Polygon'.

16.5 EVALUATION AND OUTLOOK

The rainwater management programs executed principally by ASA with governmental financing are a success story. 'We came from 1 million dead people to 1 million cisterns. In the drought of 1979 to 1982, about 1 million people in the Northeast died of starvation, that is, hunger or thirst. In the drought that lasted from 2012 to 2017, there are no records of deaths by starvation, no large migrations, no emergencies and much less looting in the cities of the hinterland' (Malvezzi, 2016). Strategic actions to control environmental degradation were carried out in partnership with communities, governments and popular movements. Thus, hundreds of people participated in simplified water management courses for food production, which aims to teach about how to care for the soil, animals and plants, thus enabling food production with better quality. These measures contributed to food security, poverty alleviation and resilience also in drought years (IRPAA, 2012). During this time, Pernambuco and Bahia state governments promulgated 'Living in Harmony with Semi-arid Climate' laws (Bahia, 2016; Pernambuco, 2013). When in the future the political situation will once again be favorable, the people of SAB will work to achieve a similar federal law.

The One Million Cisterns Program received the Future Policy Award 2017 during the 13th Conference of the Parties of the UN Convention to Combat Desertification in Ordos, China, because it 'is a participative, bottom-up way to provide water for consumption, for producing food and raising livestock in Brazil's drought-prone semi-arid region using simple rainwater collection technology. It empowers millions of the region's poorest people to be in control of their own needs, to generate income and enhance their food security' (World Future Council, 2017). In our understanding, the award may encourage other semi-arid regions of the world to carry out RWH projects, taking into account the different environmental, social, technological, economic and political factors as the FAO (2018) initiative "One million cisterns for the Sahel". Rainwater management programs are part of the sustainable development program 'Living in Harmony with the Semi-arid Climate'. According to the principle of subsidiarity, appropriate technologies supported by public funds, are carried out by families and local communities, and guarantee their participation during the implementation and maintenance. Inclusion of women is a vital ingredient for social betterment along with economic efficiency. A strong political organization of the civil society in Northeast Brazil at the local level and its networking at the state and federal level, also assist family farmers to improve their living conditions.

Now these achievements gained through public policies are at risk because of reduced support of rainwater harvesting programs by the federal government. In addition, the return of technological state-driven policies are associated with preferably large-scale interventions such as mining, energy and irrigation projects, which destroy the Caatinga, exploit natural resources, with little or no benefit to the local population. At the same time and perhaps still more serious, the changes in the climate, such as less rainfall, higher evaporation rate, longer lasting droughts and desertification create new environmental and economic conditions and challenges.

'Resilience is the capacity of a system, be it an individual, a forest, a city or an economy, to deal with change and continue to develop. It is about the capacity to use shocks and disturbances, such as a financial crisis or climate change, to spur renewal and innovative thinking. Resilience thinking embraces learning, diversity and above all, the belief that humans and nature are strongly coupled, to the point that they should be conceived as one social-ecological system' (Moberg & Simonsen, 2014). The people of SAB are aware of this.

REFERENCES

ASA (2011). *Cisternas de Plástico – Solução ou armadilha? Plastic cisterns – Solution or Trap, Recife* (in Portuguese) http://caritas.org.br/cisternas-de-plasticopvc-somos-contra/8376 [Accessed 26 July 2019].

ASA (2019). *Mapa de Technologias* (Map of Technologies), Recife https://asabrasil.org.br/mapatecnologias/ [Accessed 26 July 2019].

Bahia (2016). *Lei Estadual de Política de Convivência com o Semiárido* (State Law of Politics of Living with the Semiarid Climate), (in Portuguese) https://www.ana.gov.br/todos-os-documentos-do-portal/documentos-sre/alocacao-de-agua/oficina-escassez-hidrica/legislacao-sobre-escassez-hidrica/bahia/lei-no-13-572-2016-politica-semiarido [Accessed 26 July 2019].

Brazil, Ministry of National Integration (2005). *A nova Delimtação do Semiárido Brasileiro* (The new Delimitation of Semi-arid Brazil), Brasilia http://adcon.rn.gov.br/acervo/emater/doc/DOC000000000041645.PDF [Accessed 26 July 2019].

Brazil, Presidency (1997). 'Brazilian Water Law', Law N° 9,433, Art. 1, III, published 8 January 1997, Brasilia, Brazil (in Portuguese) http://www.planalto.gov.br/ccivil_03/leis/L9433.htm [Accessed 26 July 2019].

EMBRPA Semiárido (2019). *Dados Meteorológicos, Petrolina* (in Portuguese) http://www.cpatsa.embrapa.br:8080/servicos/dadosmet/cem-mes.html.

FAO (2018). One Million Cisterns for the Sahel, Rome www.fao.org/3/ca0882en/CA0882EN.pdf [Accessed 26 July 2019].

Girardi C. and Girardi R. (2001). *Previsão de Período de Seca para o Nordeste do Brasil (Forecast of Drought Events in Semi-arid Brazil), São José dos Campos* (in Portuguese) https://www.fundaj.gov.br/images/stories/nesa/cta.pdf [Accessed 26 July 2019].

Gnadlinger J. (2003). Rainwater Catchment and Sustainable Development in the Brazilian Semi-Arid Tropics (BSATs) – An Integrated Approach. Proceedings of the 11th International Rainwater Systems Conference, Mexico City https://irpaa.org/publicacoes/files/11thircsc.pdf [Accessed 26 July 2019].

Gnadlinger J. (2007). P1MC and P1+2, two Community Based Rainwater Harvesting Programs in Semi-Arid Brazil. Proceedings of the 13th International Rainwater Systems Conference, Sidney https://pdfs.semanticscholar.org/81c9/5a2f7a60995511892fd0ada78cf6f1c71e29.pdf [Accessed 26 July 2019].

Gnadlinger J. (2011). *A Busca da Água no Sertão*, Publicação do IRPAA, Juazeiro, 5th Edition (in Portuguese) http://www.abcmac.org.br/files/downloads/a_busca_da_agua_no_sertao_small.pdf.

Gnadlinger J. (2014). How can rainwater harvesting contribute to living with droughts and climate change in semi-arid Brazil? *Waterlines*, **33**(2). [Accessed 26 July 2019] https://www.developmentbookshelf.com/doi/pdf/10.3362/1756-3488.2014.015.

Gnadlinger J., Silva A. S. and Brito L. T. L. (2007). P1+2: *Programa Uma Terra e Duas Águas para um Semiárido sustentável*. In: *Potencialidades da água de chuva no Semiárido brasileiro*, G. F. B. Gama (Org.), Petrolina-PE: EMBRAPA Semiárido, pp. 63–77 (in Portuguese) https://www.embrapa.br/busca-de-publicacoes/-/publicacao/159651/p1--2-programa-uma-terra-e-duas-aguas-para-um-semi-arido-sustentavel [Accessed 26 July 2019].

Heijnen H. (2013). Enhancing economic resilience in North Eastern Brazil by harnessing rain. Rain Foundation, Amsterdam http://www.abcmac.org.br/files/downloads/heijnen_rainwater_harvesting_in_semiard_brazil_compressed.pdf [Accessed 26 July 2019].

IBASE, Brazilian Institute of Social and Economic Analyses (1989). *O Genocidio do Nordeste*, 1979–1983 (The Genocide in the Northeast between 1979 and 1983). Editora Mandacaru, São Paulo, Brazil.

INSA, *Instituto Nacional do Semiárido* (National Institute for the Semiarid Region) (2017). Divisão Político Administrativa do Semiárido Brasileiro, Campina Grande. PB, 2017 https://portal.insa.gov.br/images/imagens-noticias/2017/11/MAPA_POLITICO.pdf [Accessed 26 July 2019].

IRPAA (2009). *Recaatingamento – Preservação do meio ambiente em parceria com a comunidade* (Recaatingamento – Preservation of the Environment in Partnership with a Community) (in Portuguese) http://www.recaatingamento.org.br/ [Accessed 26 July 2019].

IRPAA (2012). *Seca no Semiárido?* (Drought in Semi-arid Brazil?) (in Portuguese) https://irpaa.org/publicacoes/artigos/seca-no-semiarido.pdf [Accessed 26 July 2019].

Malvezzi R. (2016). *De 1 milhão de mortos para 1 milhão de cisternas*, EcoDebate, ISSN 2446-9394, 12/12/2016, https://www.ecodebate.com.br/2016/12/12/de-1-milhao-de-mortos-para-1-milhao-de-cisternas-artigo-de-roberto-malvezzi-gogo/ [Accessed 26 July 2019].

Moberg F. and Simonsen St. H. (2014). What is Resilience? An introduction to social-ecological research. Stockholm Resilience Center, Stockholm http://www.stockholmresilience.org/download/18.10119fc11455d3c557d6d21/1398172490555/SU_SRC_whatisresilience_sidaApril2014.pdf [Accessed 26 July 2019].

Pernambuco (2013). *Lei Ordinária N° 1295/2013 Política Estadual de Convivência com o Semiárido* (Law of State Policy of Living with the Semi-arid Climate), (in Portuguese) https://issuu.com/fotoalepe/docs/2302pl01 [Accessed 26 July 2019].

Qiang Z. and Li Y. (2009). A Sustainable Way for Integrated Rural Development in the Mountainous Area in China. Proceedings of the 7th Brazilian Symposium on Rainwater Catchment and Management, Caruaru http://www.abcmac.org.br/files/simposio/7simp_zhu_paperbrazil.pdf [Accessed 26 July 2019].

Schistek H. (2010). *A Bomba d' Água Popular* (The Popular Hand Pump), IRPAA, Juazeiro (in Portuguese) http://www.irpaa.org/publicacoes/artigos/bap-artigo-revisado.pdf [Accessed 26 July 2019].

Schistek H. (2012). *O Surgimento do Barreira Trincheira* (The Origin of the Trench-like Rock Cistern). Proceedings of the 8th Brazilian Rainwater Catchment and Management Symposium, Campina Grande (in Portuguese) https://irpaa.org/publicacoes/artigos/ barreirotrincheira---s.pdf [Accessed 26 July 2019].

Schistek H. and Gnadlinger J. (2011). Technical Presentation of Self-standing Wire Mesh Cisterns built in Semi-arid Brazil. 15th IRCS Conference, Taipeh http://www.

solutionsforwater.org/wp-content/uploads/2012/02/15IRCSC_wire_mesh_cistern.pdf [Accessed 26 July 2019].

da Silva M. S. L., Mendonça C. E. S., Anjos J. B. dos, Honório A. P. M., Silva A. de S. and Brito L. T. de L. (2007). *Barragem subterrânea: água para produção de alimentos* (Subsurface Dam: Water for Food Production). In: *Potencialidades da água de chuva no Semiárido brasileiro*, G. F. B. Gama (Org.), Petrolina-PE: EMBRAPA Semiárido, pp. 121–137. https://www.embrapa.br/busca-de-publicacoes/-/publicacao/158489/barragem-subterranea-agua-para-producao-de-alimentos [Accessed 26 July 2019].

Thomas T. (2001). Choosing Rainwater Tanks for the Sertão. Proceedings of the 3rd Brazilian Rainwater Catchment and Management Symposium, Campina Grande, Brazil http://www.abcmac.org.br/files/simposio/3simp_terrytomas_choosingrainwater.pdf [Accessed 26 July 2019].

World Future Council, Future Policy Award (2017). Brazil Cisterns Program empowers millions of people https://www.worldfuturecouncil.org/p/2017-desertification/ [Accessed 26 July 2019].

Chapter 17

Every drip counts: Confusion of cause with effect in the climate debate

M. Schmidt

Technische Universitat Berlin, A 59, Building Technology Research Group, Strasse des 17 Juni 152, Berlin, Germany
marco.schmidt@tu-berlin.de

Keywords: artificial lake, evaporation, precipitation, radiation, rainwater harvesting, water

17.1 BACKGROUND

The main component of the water cycle has been up to now almost totally ignored: evaporation. As evaporation is responsible for all precipitation, it is a key driver for the global, and as well local, climate. Figure 17.1 shows the energy budget on the surface of the earth as mean of one square meter worldwide. Incoming radiation is converted into four components: (a) reflection and (b) longwave emission, as part of direct radiation components; (c) sensible heat; and (d) evaporation as converted energy components, also summarised as net radiation.

The global energy budget is dominated by evaporation and condensation. While reflection and longwave emissions represent 7% and 38% of incoming shortwave radiation, respectively, the largest component represents evaporation at 43%. Furthermore, evaporation reduces longwave thermal radiation due to the decrease in surface temperatures.

A reduction in evaporation as e.g. result of urbanization processes is shown in Figure 17.2. Urbanization results in huge changes to the small water cycle. While

 International Rainwater Catchment Systems Experiences: Towards Water Security
Editor(s): José Arturo Gleason Espíndola, César Augusto Casiano Flores, Raul Pacheco-Vega and Margarita Rosa Pacheco Montes
doi: 10.2166/9781789060584_0221

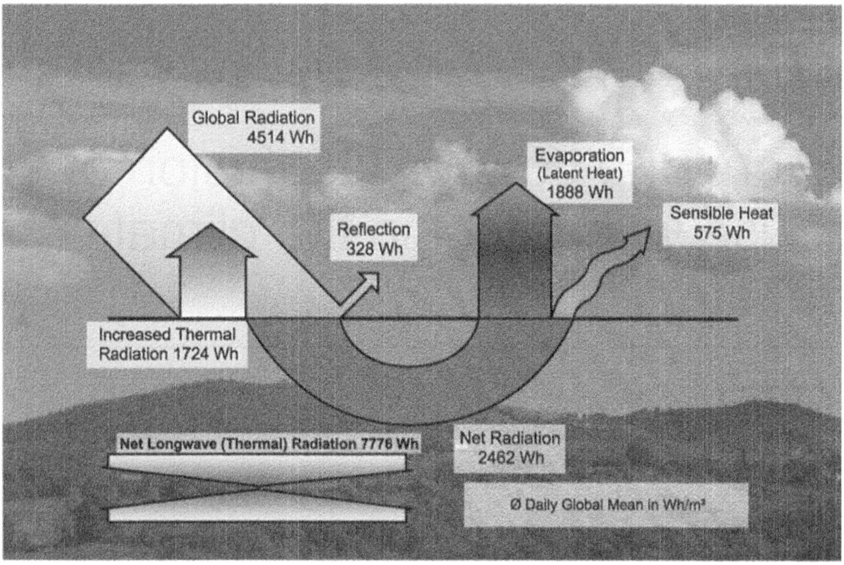

Figure 17.1 Average global daily radiation budget of one square meter worldwide (Schmidt, 2010). Energy data based on www.physicalgeography.net.

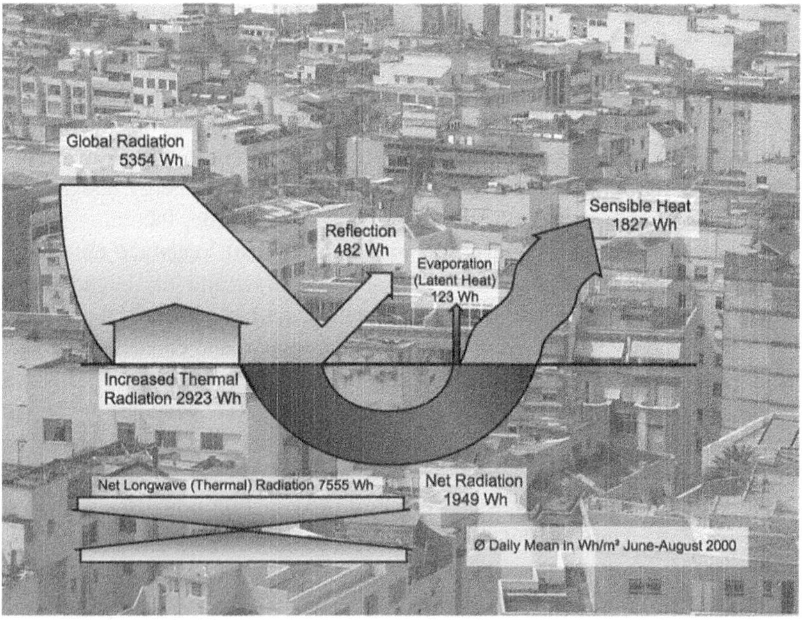

Figure 17.2 Radiation budget of a black asphalt roof as an example for urban radiation changes (Schmidt, 2005).

rainwater is funnelled into sewer systems, surfaces absorb and re-radiate solar irradiation. As a result, air temperatures inside buildings also rise, leading to greater energy consumption from air conditioning. This worsens the situation of the urban heat island effect by a release of additional heat due to the use of electricity for cooling (Schmidt, 2003). To exemplify radiation in urban areas, Figure 17.2 illustrates the radiation budget of a black asphalt roof in the summer months in Berlin. While rainwater disappears into sewer systems, most of the net radiation from the urban setting is converted to sensible heat rather than evaporation. Higher surface temperatures also increase thermal radiation.

The option to green buildings is a logical solution to create healthy and sustainable air temperatures in cities and improve microclimate. Vegetation on and around buildings converts solar radiation into latent heat by evapotranspiration. A further option is rainwater harvesting from roofs and paved surfaces to be used for irrigation. A third option is rainwater use in air conditioners to provide evaporative cooling. This avoids the use of conventional compression cold. While conventional systems are based on the use of electricity and therefore emit heat, evaporation results in real cold and finally precipitation.

While urbanization increases at a global rate of 150 km^2 daily, reduced evaporation results in reduced precipitation and increase in temperatures.

A new water paradigm was established by Kravčík *et al.* (2007). While evaporation in the 'virtual water' concept is defined as a loss, after Kravcik it is expressed as source of precipitation. Local precipitation rates are dominated by the small water cycle, evaporation on land and resulting precipitation on land (Figure 17.3).

For the area Berlin/Brandenburg, Germany, a reduction in evaporation of 1 m^3 due to unsustainable land use, results in a reduction of precipitation on land of 5 m^3. In the catchment area of Berlin/Brandenburg, about 80% of precipitation is converted to evaporation, while groundwater recharge and runoff together represent 20%. Urban areas are characterized by completely paved or semi-permeable surfaces with little to no vegetation. Semi-permeable surfaces allow much higher groundwater recharge compared with naturally vegetated areas (Schmidt, 2005), as they over-compensate for infiltration with reference to completely paved surfaces. Therefore, in the interest of effective environmental care taking, the provision for evaporation rather than infiltration needs to become a primary task.

Implementing the new climate paradigm with a focus on evaporation at a local level requires rethinking urban planning and water management infrastructures. With regard to the urban heat island effect and the issue of global warming, urban planners, architects and landscapers need to consider the natural water cycle, including evaporation, condensation and precipitation (Milosovicova, 2010). The conventional management of water discharge, which was implemented for over a hundred years, nowadays bears disastrous environmental

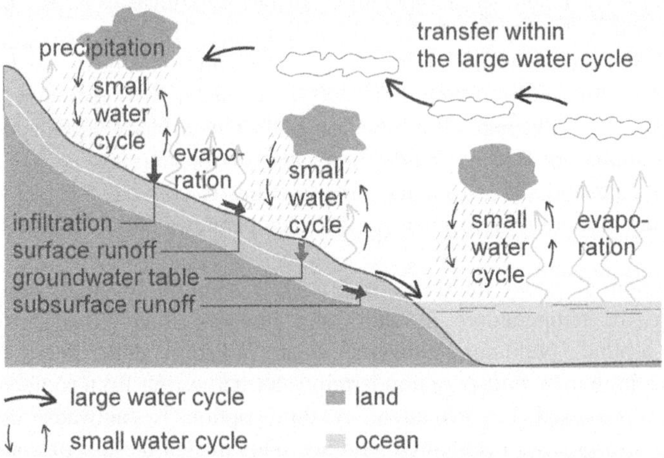

Figure 17.3 Large (ocean–land) and small (land–land and ocean–ocean) water cycles (Kravčík *et al.* 2007).

effects on both surface water quality and on the climate. More recently, rainwater infiltration has been a popular strategy in Germany. However, in spite of the great benefit of preventing negative impacts to surface waters, infiltration does not fully rectify the natural water cycle. Urban areas are not characterized by reduced infiltration rates (SenStadt, 2013). The missing hydrological component is evaporation.

Through urban design strategies such as green facades, green roofs and stormwater ponds, rainwater management strategies play supportive roles as adaptation and mitigation strategies against the urban heat island effect and global warming. By applying the new paradigm small local water cycles are restored (see www.rainforclimate.com). Harvesting rainwater for evaporation should be a first priority in urban areas. Several projects have been established in Berlin with a focus on rainwater harvesting and evaporation, of which two are presented here, Potsdamer Platz and Adlershof Physics Building.

17.2 POTSDAMER PLATZ

A large project supporting the new water paradigm is the former DaimlerChrysler-Area at Potsdamer Platz in Berlin (https://potsdamerplatz.de/en/sustainability/). From 1996–1998 the largest construction site in Europe was built under very strict stormwater management regulations. In order to avoid overloading the existing combined sewerage system in central Berlin, the building permit issued by city council stated that the new complex would drain runoff at a rate of no more than 3 l/sec/ha, or 1% of flows during storm events. To comply with this regulation, the Atelier Dreiseitl (www.dreiseitl.com) and landscape architect Daniel Roehr in cooperation with Technische Universität Berlin implemented the following techniques for the management of 23 000 m^3 precipitation which falls annually on this building site:

- extensive and intensive green roofs on all of the 19 buildings;
- collection of roof runoff for toilet flushing and plant irrigation;
- an artificial lake for rainwater retention and evaporation.

Since infiltration was not possible at this site, the basis for the rainwater management concept involved rainwater harvesting for toilet flushing and evaporation by green roofs and an urban lake as a retention pond. Three cisterns providing 2550 m^3 storage capacity correspond directly to 12% of the annual precipitation of the catchment area. The artificial lake (Figure 17.4), covering a total area of 13,000 m^2, can fluctuate its levels by 30 cm, which corresponds to an additional storage capacity of 11% of the annual precipitation. The water is cleaned and filtered through artificial filtering systems and additionally by a constructed wetland of 1900 m^2 which is planted mainly with Phragmites. The resulting water quality, as well as stormwater issues, has proven that this large rainwater system has performed very well for the last 20 years of operation.

Figure 17.4 Urban lake, supplied with roof-runoff at Potsdamer Platz. (*Source*: Author's own).

17.3 ADLERSHOF PHYSICS BUILDING

Another innovative project providing the new climate and water paradigm on evaporation is the Institute of Physics in Berlin-Adlershof (Figure 17.5). The building is located in a research and office facility featuring several measures of sustainable architecture. It was designed by the architects Georg Augustin and Ute Frank (Berlin) following an architectural competition held in 1997. Rainwater is used to supply a façade greening system and central air-conditioning systems with evaporative exhaust air cooling. The water is harvested from the roofs and stored in five cisterns.

Research elaborating on the performance of the building is carried out by Technische Universität Berlin and funded by the Berlin Senate of Urban Development, the Federal Ministry for Science and Technology and the Federal Ministry for Environment. The project includes permanent monitoring of the water consumption of different plant species and eight air conditioning units. Continuous monitoring has been carried out since 2004.

The façade greening system is evaluated to determine the importance of evapotranspiration and shading on the overall energy performance of the building, including temperature and radiation measurements. Data collected from this project is used to calibrate simulations that are designed to predict performance and benefits in range of different climatic conditions. This work will inform the design of future projects (SenStadt, 2010).

About 300 parameters are electronically harvested every minute. Primary systematic evaluation is based on the water parameters and their relation to energy dissipation. Additionally, 20 plant species and their requirements for maintenance (fertilization, plant protection) are monitored. Seven long-wave,

Figure 17.5 Façade greening system (left), artificial rainwater pond combined with trough infiltration for stormwater management (right). (*Source*: Author's own).

short-wave and infrared sensors monitor the radiation concerning shading and reflection for a conventional façade with sunblinds and the greened façade. The building is not connected to stormwater sewers, reflecting one of the main goals of this decentralized system of rainwater retention and harvesting. Stormwater events from heavy rainfall are managed with an overflow into a small constructed pond in one of the courtyards, from which the water can evaporate or drain into the ground. To protect the quality of groundwater, this drainage is only allowed through surface areas covered with vegetation. Some of the roof surfaces are also extensively greened to assist in retaining and treating stormwater.

REFERENCES

Kravčík M., Pokorný J., Kohutiar J., Kováč M. and Tóth E. (2007). *Water for the recovery of the climate – A New Water Paradigm:* Retrieved on June 2019 from http://www.waterparadigm.org.

Milosovicova J. (2010). 'Climate Sensitive Urban Design: Responding to Future Heatwaves. Case Study Berlin – Heidestrasse/ Europacity'. Master Thesis, TU Berlin, 94. http://www.jm-urbandesign.com/images/Thesis%20document.pdf.

Schmidt M. (2003). Energy saving strategies through the greening of buildings. Proceedings of Rio3, World Energy and Climate Event, Rio de Janeiro, Brazil, Retrieved on May 2019 from http://www.rio3.com.

Schmidt M. (2005). The interaction between water and energy of greened roofs. Proceedings World Green Roof Congress, Basel, Switzerland, 15–16th September, 2005.

Schmidt M. (2010). Ecological design for water and climate mitigation in contemporary urban living. *Int. J. Water*, **5**(4), 337–352.

SenStadt. (2010). Rainwater management concepts: Greening buildings, cooling buildings. *Planning, Construction, Operation and Maintenance Guidelines.* Senatsverwaltung für Stadtentwicklung, Berlin, Germany, Retrieved on June 2019 from www.gebaeudekuehlung.de.

SenStadt. (2013). Environmental Information Systems, Berlin Digital Environmental Atlas, Map 02.13 Surface Runoff, Percolation, Total Runoff and Evaporation from Precipitation: Retrieved on June 2019 from http://www.stadtentwicklung.berlin.de/umwelt/umweltatlas/eic213.htm.

Chapter 18

The Calabash Cistern 5000 L in Africa

P. Akkerman

Independent consultant Netherlands and Guinea-Bissau
info@degevuldewaterkruik.nl, www.cleanwaterhealthyvillage.com

Keywords: Africa, calabash cistern, drinking water, rainwater, tank, water

18.1 INTRODUCTION

Skillful farmers and local masons have constructed thousands of 5000 L cisterns in remote villages and on far away islands, in more than ten African countries (Figure 18.1). The cisterns are built at low cost, approximately €240 in 2019, and are made with locally available tools and materials. The idea is to collect rainwater and store it close to the kitchens of farmer's wives. The basic issue is the existential necessity for small peasant families to have safe drinking water. The driving force is to relieve millions of women and girls from their heavy burden of having to carry drinking water, every day and for many hours. This chapter tells the story of the Calabash Cistern in short.

18.2 THE DESIGN OF THE CALABASH CISTERN (FIGURE 18.2)

18.2.1 Development

It has taken nine years to develop the design of the Calabash Cistern. In this story we call our project 'The Calabash Project'. The original name is CLEAN WATER – HEALTHY VILLAGE.

Editor(s): José Arturo Gleason Espíndola, César Augusto Casiano Flores, Raul Pacheco-Vega
and Margarita Rosa Pacheco Montes
doi: 10.2166/9781789060584_0229

Figure 18.1 Calabash cistern 5000 L. (*Source*: Author's own).

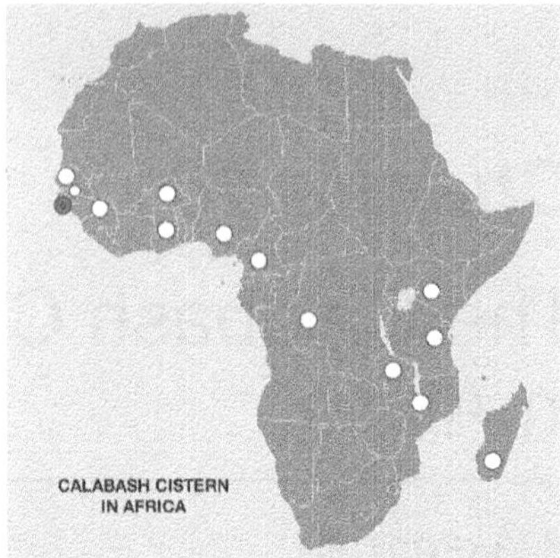

Figure 18.2 Calabash cistern spread over Africa. (*Source*: Author's own).

To reach the Calabash design and the construction technology with an outside mould, there were eight steps in the development.

Since the start of our project in 2005 we have improved the technology and the shape of our cisterns. All the experiments were performed in one of our three small water centres and in the African villages. Once a year we invite our best trainers and masons for a training meeting and an exchange of new ideas. Our development is a typical African development. Our manual for the construction of the Calabash Cistern is published in three languages. Our knowhow is available for free. RWH-consultant Hans Hartung visited and evaluated our project. In his report he writes: The Calabash Cistern is a development of its own.

18.2.2 The 5000 L Calabash Cistern

The overall shape is a vertical cylinder with a round-shaped bottom and a slightly conical roof. The material used is ferrocement. The cylinder has an inner diameter of 200 cm. The round shape of the connection between the bottom and the wall has a radius of about 40 cm. The height from bottom to the edge of the roof is between 1.70 and 1.80 m depending on the size of the mould.

We chose ferrocement as the material to construct the water cisterns, because we want the cisterns to be constructed by local craftsmen in rural areas. Cement is a very good material to store water. It is cheaper than stainless steel and more resistant than plastic. It can be constructed on the spot using hand tools. Cement is available almost everywhere in rural areas. A leaking wall is reparable. The life span can be 50 years, depending on maintenance.

18.2.3 Shape

The round-shaped connection between bottom and wall is essential in the design. The round shape looks like the shape of a water drop and is related to the

Figure 18.3 Preparing the hollow part of the mould. (*Source*: Author's own).

Figure 18.4 Plastering the inside of the mould. (*Source*: Author's own).

traditional African water jar. Being the designer, I felt very happy when I found the technical solution to make a globe or round shape for the cistern. I remember the moment when the idea struck me: 'In Guinea-Bissau I was staring out of the window of a hot bumpy bus driving through a Fula-tribe village with its round houses made of clay blocks. All the time I had been trying to imagine how to construct a suitably shaped water cistern. It had to be a round, comfortable and pleasant shape for water storage. All at once I got an idea: why don't I use a round outside mould instead of an inside mould?' An outside mould gives the possibility to shape a hollow curve in wet sand by hand on the inside of the cylindrical wall of blocks (Figure 18.3, 18.4 & 18.5). The method is similar to the work of an African smith, who is casting melted aluminium in a mould to make a globe shaped marmite pot. When the pot is finished, the smith takes off the mould from the outside (Figure 18.6).

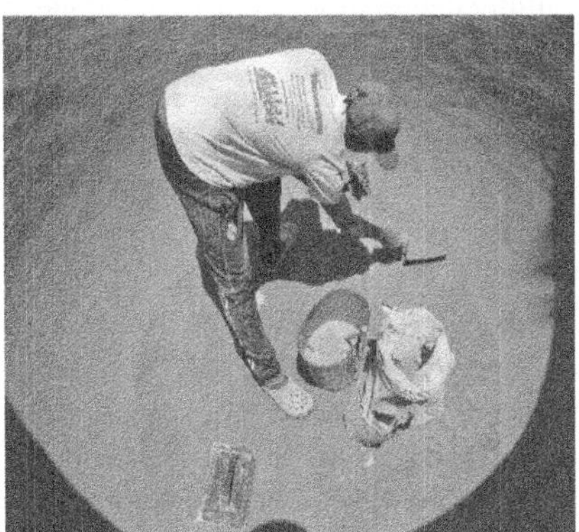

Figure 18.5 Finishing inside a calabash cistern. (*Source*: Author's own).

Figure 18.6 Finishing the outside. (*Source*: Author's own).

Our project already worked with an inside mould of clay blocks. Now we changed the method. Together with Julio, my best mason, we constructed the new round-shaped cistern and we called it the Calabash Cistern.

18.2.4 Advantages

An important advantage we found on construction is that the round shape avoids the sharp edge where the bottom of the cistern meets the wall. The mechanical tensions are spread over the surface. In practice we do not see any cracks in the body of the Calabash Cistern. Another advantage is that the new shape saves one bag of cement. Also, most of the plastering can be done in the shade inside the cistern, which makes the quality of the cistern better.

18.3 THE CALABASH PROJECT IN GUINEA-BISSAU

18.3.1 How it started

The Calabash Project started in a small village and it walked over the narrow roads and footpaths of friendship and families to the next families and the next villages, like a cascade. Several years later big leaps were made to extend the reach of the projects in different countries. On the African family and friendship roads, one sometimes meets unexpected problems, but African families have proved to be very powerful. In every village we entered, the project leader was welcomed by motivated and skilful farmers. The pillars of the project proved to be: (1) the skilful African village; (2) the unbearable burden of women carrying so much water; (3) the local invitation.

No time for school.

I worked and lived in a village called Bedanda for almost four years (1987–91) and I earned my living at a rural vocational training centre as a mechanical engineer and teacher. My personal motivation was to experience more about the cooperation and friendship with African farmers. I had met them on my long holiday walks. They were very vital, skilful and hospitable people. I decided to go and work in an African rural area and I was lucky enough to get a job in Bedanda, a village of animist Balanta rice farmers.

A Balanta village has many spirits. When, the first monsoon rain arrives in Bedanda, the rice farmer awakes from his winter sleep and the rice-spirit affects the whole village population. People start their yearly dance of ploughing, sowing, harvesting, etc. Men and women, boys and girls are fully involved. Rain brings happiness. I met my best friend Bicosse Nandafa and he introduced me to many aspects of the Balanta society: birth, life and death. In this period, we started several projects to support children. The awareness of a drinking water project came later. Bicosse introduced me to working and communicating with the biggest tribe in Guinea-Bissau.

18.3.2 The 1st pillar of the project: The African village

The farmers of Bedanda invited me to work together with them on the rice fields with their very special hand plough, called the aradu. The Balanta farmers have a well organised agricultural system. They are skilful, hardworking self-sufficient farmers in a complicated society. During a few seasons, I was almost one of them, both in working with them and taking part in their parties and ceremonies. They welcomed me and I learned the language. I was amazed about their extraordinary effort to plough thousands of acres by hand. (The Dutch Agricultural University Wageningen did local research). Balanta men and women maintain an extended network between independent villages for exchange of labour, marriage, education of children and inauguration. Because of their skills, I felt sure that I could start a project directly with the village population instead of a complicated cooperation with an instable government.

18.3.3 The 2nd pillar (and driver): A private water cistern

Imagine, a private water cistern belonging to your own household! This was the driver of the project. Balanta women suffer because of bad drinking water, for the women it is a hard working life.

In the coastal area of Guinea-Bissau there is abundant rainfall during five months, then not a single drop of rain for seven months. The groundwater is saline because of intrusion of seawater, so the wells provide polluted water. In the dry season women and daughters have a daily walk of several hours to fetch and carry clean water from inland wells. (Figure 18.7 & 18.8) Women do not like living in the village of Bedanda, because of these heavy duties. For men it is difficult to find a wife who is willing to marry them.

Women and men want to store drinking water safely at home in large amounts, just as they store their rice at home, inside the house. A village well, a village pump and a village tap belong to the community. Something for common use, unfortunately, will not be maintained and suffer 'the tragedy of commons'.

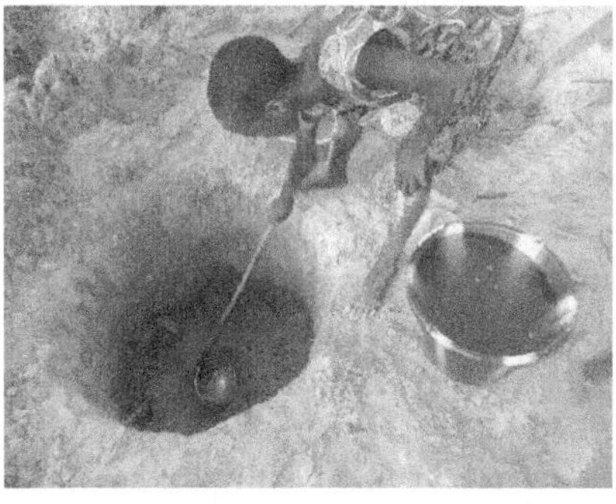

Figure 18.7 Drinking water taken from a shallow well. (*Source*: Author's own).

Figure 18.8 Long walks to transport drinking water. (*Source*: Author's own).

18.3.4 The 3rd pillar: The initiative

On behalf of the village population, my friend Bicosse asked me for help to find a solution for the drinking water problem in his village Bedanda. Very importantly, he asked me and not the other way round. In the village, Bicosse was an adult and inaugurated man, aged around 40 years. He had an important position in the community of men. His younger brother was a mason in the capital Bissau. I found some financial support from my friends in the Netherlands and some technical information about storing rainwater. We started right away like amateurs. In the same year, 2005, we constructed three 5000 L cisterns in the ground. They were filled up by rainwater from the roof. One of the three cisterns was good and leak free. Even after six months, the taste of the water was very good. Nobody had seen anything like it in the village. Big local news.

18.3.5 The start of the Calabash project was booming

The second year I brought an experienced mason from the Netherlands to train a team of local men. We constructed ten underground cisterns. The 3rd year another training and we started to construct the cisterns above the ground, which is easier to control, more reliable and reparable. We constructed 50 cisterns above the ground. The 4th year we constructed 100 cisterns. Every household wanted to have water in stock, the neighbours, the brothers and the sisters, the neighbouring villages.

We started to ask a contribution in money. A cistern is not for free, it is not an easy present, ('You can just put it there!'). A cistern needs a serious owner, who will think and care about it and lock it at night!

18.3.6 The Calabash Project sails to the islands of Guinea-Bissau

South West Guinea-Bissau consists of many fertile islands and peninsulas. Around the islands there are broad salt water sea-arms, estuaries and rivers. A tidal landscape, originally covered with mangroves. It is the biggest sea area with coastal islands in West Africa with many birds. The rice farmers know how to cultivate mangroves into rice fields. Families have settled there because of the fertile soil and rich fishing grounds. They suffer but accept bad and saline drinking water, they have little knowledge of the cause of water-related illnesses.

Through friendship and family connections, my friend Bicosse and I visited the isolated and beautiful islands. We could only sail there in a well carved, hand-powered canoe, made out of the trunk of a 'canoe-tree'. Welcoming inhabitants informed us about the drinking water problem, which was even more serious than on the mainland. One of the islands is called Widekea. It means 'no water'.

Figure 18.9 Polluted water. (*Source*: Author's own).

Figure 18.10 Clean water at home. (*Source*: Author's own).

As a result of its isolation, healthcare was very poor. It took four or five hours to paddle pregnant women to the hospital in Catio. Many times the baby died half way. The mortality rate of children under the age of five was 20%. Local nurses told us that mortality depends on a combination of illnesses, which make a child weaker and weaker. Illnesses like diarrhoea, malaria, infections, etc. The animist population believes that illness and death are not caused by bacteria or dirty water, but by ghosts and curses (Figure 18.9). The following story about the influence of safe water (Figure 18.10) on the perception of cholera is striking.

18.3.7 Clean water changes opinion about cholera

In 2014, my wife (she is a family doctor) and I visited the very fertile island of Catunco. The project had been constructing Calabash Cisterns since 2010 and we wanted to visit the population and to inspect the results. There had just been a cholera epidemic in 2013. 20 people had died and half of the population of 3500 souls had fled for fear of infection. We watched a very big meeting on the football field with hundreds of people, men, women, old and young in different groups together. We did not want to disturb them. Our host, a teacher, explained that the goal of the meeting was to find out who was responsible for the curse. An explanation and a discussion between my wife, the doctor, and the teacher did not work out. The teacher assured us that only the old, male village leaders have the power to solve the problems and not the doctors from the universities.

In 2018 two project leaders of Catunco came and visited the project centre in Catio. We were sitting in my room. It was a conversation about money, so I closed the door. They told me that since the Calabash project started 10 years ago, there had not been a cholera epidemic on the islands for five years now. They shared with me that their families and friends had started to believe that a cholera outbreak depends on water and that the illness is not the result of a curse. They confessed that until five years ago people were killed in the night or expelled from the island because of the suspicion of black magic.

18.3.8 Transport and salty sand

As the Calabash Cistern is constructed on the islands, the project has to deal with transport of the materials. One Calabash Cistern needs eight bags of cement, 10 kg of steel wire, 8 m of chicken mesh and a tap. One hand-driven canoe can easily transport two masons, the materials and the tools for five cisterns. Nowadays the project has its own motorized canoe, big enough to transport the material for 10 Calabash Cisterns.

The first cisterns on the islands were made with beautiful white sand, straight from the beach. In the first year these cisterns worked well. In the second year most started leaking. The sea salt in the sand diluted and made the wall permeable. The lesson was to start washing the sand thoroughly in the rainy season.

18.3.9 Another change

In 2015 Nobel Peace Prize winner, José Ramos Horta was working in Guinea-Bissau as a representative of UN. After several requests and a long period of waiting, he decided to visit our project on four of the islands. He showed enthusiasm about the cisterns and he talked with the population about their needs. They asked him to get help with transport and Mr. Ramos Horta offered each of the four islands a motorised canoe, so that the islands would have better transport to the hospital on the mainland.

18.4 THE FIRST SUCCESSFUL INTERNATIONAL TRAINING IN DR CONGO

18.4.1 The importance of the manager

During our first successful training, we learned the importance of a good manager. In two earlier international projects, we had trained very good masons. Yet, in spite of training, the projects failed. However, in Kinshasa we not only met masons, we also trained the local project coordinator or manager, Ir. Roger Mbumbu. Ir. Roger has his own team of masons and for us he has become a role model as a local project coordinator.

In the Netherlands we have hundreds of private development organisations working for small and poor communities in Africa, Asia, etc. They support hospitals, they build schools, they finance pumps for irrigation, etc. Our foundation is one of them, Foundation Friends Holten Bedanda - friendship between a Dutch and an African village.

To bring all these private foundations together, we have an umbrella organisation, called Partin. Once a year Partin organises a congress for all her members. In 2013 they invited me to give a presentation about our drinking water project and the cisterns. There I met Gustave, who is the coordinator of a project in DR Congo. Gustave was very enthusiastic about our cisterns, we made a plan and his organisation invited our trainers to train their masons in the town Menkao not far from Kinshasa. (The groundwater in the area is 200 m deep.) They found funds to pay for the training and the tickets for the trainers from Guinea-Bissau. African trainers train African masons! (Figure 18.11) My role was to supervise the training, to be an interpreter and to work with the local manager. So I met Ir. Roger Mbumbu. He invited me to a meeting with the chief and representatives of the town. Roger organised all requirements and we agreed about future cooperation and some support from our side. Since 2014 Ir. Roger has constructed about 100 Calabash Cisterns for 100 families in Menkao. After some years Ir. Roger and Gustave succeeded in raising their own funds.

Figure 18.11 Certificates for participants, DR Congo 2014. (*Source*: Author's own).

In 2019 a Tanzanian project invited Roger to be a trainer in Tanzania. It was the first time in his life he travelled that far. He trained his Tanzanian colleagues using the Swahili language, a language he spoke because he was born in East Congo. Roger's African horizon was opened.

This is only one example of how Partin supports us. The umbrella organisation linked us with projects of at least ten African countries in West and East Africa.

18.4.2 Menkao town

Menkao town has about 35,000 inhabitants and has a history with drinking water. In the middle of the town is an impressive water castle with a 250 m deep borehole. Beautiful taps in different quarters of the town. Around 500,000 Euros was financed by the EU. After half a year, the pump broke. No money and no expert to repair.

Ours is a very different approach. For the same amount, we could have constructed more than 1000 Calabash Cisterns. Even if 10% of the cisterns had a problem, that would still be 900 working Cisterns. And the broken cisterns could be repaired by local masons.

18.4.3 Calabash Project becomes specialized in training

First we became specialized in the construction of the Calabash Cistern. After that we became specialized in the training of masons and managers. There is no need for us to grow, no big centres, no offices and cars. We like to stay small and basic and keep in touch with the African farmer. Don't forget: our trainers and masons are as famous as local football stars.

18.5 OTHER REMARKABLE RESULTS OF THE INTERNATIONAL TRAINING

18.5.1 CBR Effata

CBR Effata is a community-based rehabilitation program in Nigeria and in Guinea-Bissau. One of the activities is to support deaf children. They are brought together from different villages and trained in using sign language; the deaf boys and girls go to school and get a vocational training for a job. Several boys chose to be masons and construct Calabash Cisterns. They were trained by the trainers of the Calabash Project. Now they have their own teams for the Calabash Cisterns and every year they construct about 120 cisterns in Nigeria from their own Effata funds.

18.5.2 Reforestation Project, Senegal

For a few years the Calabash Project has been closely cooperating with the International Rainwater Harvesting Alliance (IRHA). Han Heijnen, president of IRHA invited us to participate in this Reforestation Project. In 2019 our trainers from Guinea-Bissau trained six Senegalese masons and a manager (Figure 18.12). They are now able to continue their project independently.

Together we created safe drinking water possibilities for peasant families by means of the Calabash Cisterns. Our trainers participated in many more African projects, and also in Panama and Mexico (Figure 18.13).

The training is supported by a well illustrated manual of 40 pages. It takes six days to construct a Calabash Cistern. A team of three masons can construct three cisterns together in one week by working parallel.

Figure 18.12 Participants Senegal 2019. (*Source*: Author's own).

Figure 18.13 Indigenous people in North Mexico. (*Source*: Isla Urbana, 2018)

18.5.3 The manual (Figure 18.14)

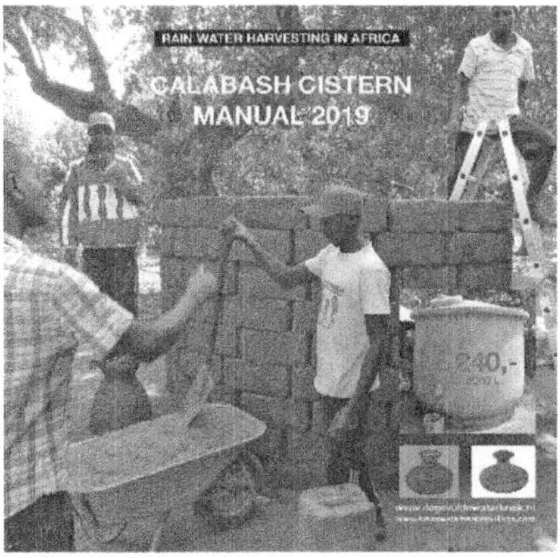

Figure 18.14 Calabash Cistern manual 2019. (*Source*: Author's own).

The illustrated manual of 40 pages leads the managers and the masons through the six-day construction period, day by day. Many years of experience have been collected in the manual. It is a supporting guideline and standard. Harry Chaplin director of Tatirano Social Enterprise in Madagascar studied the manual and was able to construct the Calabash without further training. Though this is possible, the great advantage of a training is that participants will teach each other many small tricks and details, that are not all in the manual. Information about our method of training is found on our website: www.cleanwaterhealthyvillage.com

18.5.4 The PVC tank (Figure 18.15)

The PVC tank is a competitor to the Calibash Cistern. Many thousands of plastic tanks are used in urban centres such as Dakar, Dar Es Salaam, Nairobi, etc.

In Ghana masons have constructed Calabash Cisterns in towns. Cement cisterns are cheaper, more durable and their production supports the local economy. The water in a cement cistern will stay cooler than in a plastic one. Cisterns are heavy, but can be transported with a bulldozer (Figure 18.16).

Figure 18.15 PVC tanks in Dar Es Salaam. (*Source*: Author's own).

Figure 18.16 Town transport of a 6000 L Calabash Cistern. (*Source*: Author's own).

18.6 SOME CONCLUSIONS AND OUTLOOK

It is not so difficult to construct the Calabash Cistern. It can be learned by skilful men, women and by normal masons. The discipline regarding the quality of mortar, reinforcement, wall thickness, keeping the cement wet, etc. is important, but not difficult. The Calabash Cistern is cheap and in many villages there is somebody who can construct and repair them. Calabash Cistern can reach remote households and is suitable for small projects and self help.

Detailed information cannot be explained in this short chapter. Please find the manual and more information on our website: www.cleanwaterhealthyvillage.com.

The 10,000 L Calabash Cistern is also generally constructed at schools and enterprises.

The 5,000 L Calabash Cistern is a practical volume for families. If more volume is required, then two Calabash Cisterns can be constructed.

Industrial growth in urban centres is possible like that of the production of Thai Jars.

In Brazil I visited PIMC, Program for 1 Million Cisterns, in rural areas and was very impressed with the work there.

I wish all RWH practitioners good luck and would be pleased to hear from anyone who has constructed or uses the Calabash Cistern. RWH is like singing in the rain.

Index

IWA Publishing's authorised EU representative for General Product Safety Regulations is Diane D'Arras, 15 rue Duret, 75116 Paris, France, e-mail: safety@iwap.co.uk.

Printed and bound by CPI Group (UK) Ltd, Croydon, CR0 4YY

16/06/2026

02140365-0001